He Fell a Cheerful Sacrifice to His Country's Glorious Cause

General William Woodford of Virginia, Revolutionary War Patriot

Michael Cecere

HERITAGE BOOKS
2019

HERITAGE BOOKS
AN IMPRINT OF HERITAGE BOOKS, INC.

Books, CDs, and more—Worldwide

For our listing of thousands of titles see our website at
www.HeritageBooks.com

Published 2019 by
HERITAGE BOOKS, INC.
Publishing Division
5810 Ruatan Street
Berwyn Heights, Md. 20740

Copyright © 2019 Michael Cecere

All rights reserved. No part of this book may be reproduced or transmitted in any form or by any means, electronic or mechanical, including photocopying, recording or by any information storage and retrieval system without written permission from the author, except for the inclusion of brief quotations in a review.

International Standard Book Number
Paperbound: 978-0-7884-5873-6

Heritage Books by Michael Cecere:

*A Good and Valuable Officer:
Daniel Morgan in the Revolutionary War*

*A Universal Appearance of War:
The Revolutionary War in Virginia, 1775–1781*

*An Officer of Very Extraordinary Merit:
Charles Porterfield and the American War for Independence, 1775–1780*

Captain Thomas Posey and the 7th Virginia Regiment

*Cast Off the British Yoke:
The Old Dominion and American Independence, 1763–1776*

*Great Things are Expected from the Virginians:
Virginia in the American Revolution*

*He Fell a Cheerful Sacrifice to His Country's Glorious Cause:
General William Woodford of Virginia, Revolutionary War Patriot*

*In This Time of Extreme Danger:
Northern Virginia in the American Revolution*

*Second to No Man but the Commander in Chief, Hugh Mercer:
American Patriot*

*They Are Indeed a Very Useful Corps:
American Riflemen in the Revolutionary War*

*They Behaved Like Soldiers:
Captain John Chilton and the Third Virginia Regiment, 1775–1778*

*To Hazard Our Own Security:
Maine's Role in the American Revolution*

*Wedded to My Sword:
The Revolutionary War Service of Light Horse Harry Lee*

Contents

Ch. 1   "I Thought it my Duty..." 1734-1774..........1

Ch. 2   "I Highly Approve of Your Appointment." 1774-1775..………..……..23

Ch. 3   "Americans will die, or be free!" Fall 1775..…………......................41

Ch. 4   "A Second Bunker's Hill Affair, in Miniature." Winter 1775………..…………57

Ch. 5   "I Would Wish to Keep up the Greatest Harmony Between Us." 1776 ..…………..83

Ch. 6   "I Must Request Your Permission To Retire." 1776…………………………..103

Ch. 7   "The Rebels Disputed [the ground] with Great Spirit, Particularly Their Officers." 1777…………………..125

Ch. 8 "When Rank is Once Given…the Party In Possession of it in Most Cases is Unwilling to Give It Up." 1777-1778……....151

Ch. 9 "I am against risquing a Genl. Attack."….…167
1778

Ch. 10 "For my Part I would Wish Never to Part with Him." 1778-1779…………………185

Ch. 11 "We arrived…to the great joy of the Garrison." 1780……....…………………….213

Bibliography……………………………………….241

Index………………………...…………….…….…..253

Acknowledgements

As with all of my books, I'd like to thank my friends in the Revolutionary War reenacting hobby for the continued support and encouragement. Debbie Riley with Heritage Books was immensely helpful with preparing the book for publication and a true godsend with the maps. I also want to thank the Simpson Library at the University of Mary Washington, the Rockefeller Library at Colonial Williamsburg, the Swem Library at the College of William and Mary, and the research library at the Jamestown--Yorktown Foundation. All of these wonderful facilities provided me with the resources necessary to write this book.

Chapter One

"*I Thought it my Duty...*"

1734 – 1774

William Woodford was born on October 6, 1734 in Caroline County Virginia, the eldest of five children.[1] His father, Major William Woodford, emigrated from England to Virginia in the early 1700's and settled on the northern edge of Caroline County near the Rappahannock River.[2] Major Woodford apparently served in the British army prior to his arrival in Virginia and rose to prominence in Caroline County as a merchant and planter.[3] He served as a justice of the peace and county sheriff for many years and established a prosperous plantation along the south side of the Rappahannock River which he named Windsor.[4] No trace of the residence or its outbuildings exists today, but it is believed that the bulk of the property rests within Fort A. P. Hill, a modern day U.S. army base situated mid-way between Port Royal and Fredericksburg.[5] Both towns were thriving colonial ports in the 18th century.

---

[1] Catesby Willis Stewart, *The Life of Brigadier General William Woodford of the American Revolution*, Vol. 1, (Richmond, VA: Whitten & Shepperson, 1973), 70
[2] Ibid., 19
[3] Ibid., 19-20, 49-50
[4] Ibid., 50, 20
[5] Ibid., 20

Map of Caroline County

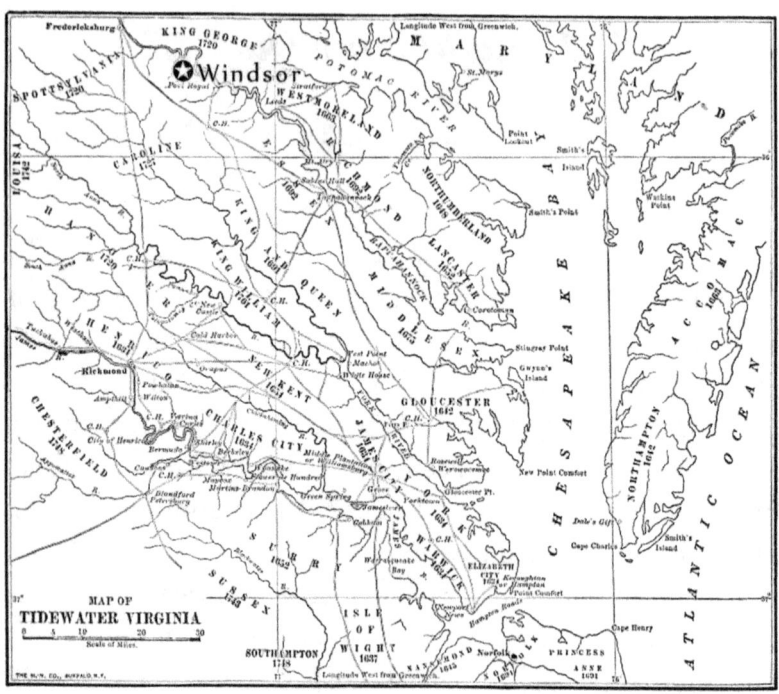

Although Major Woodford enjoyed financial good fortune in Virginia, he suffered personal loss with the deaths of his first and second wife. His family situation brightened considerably with his third and final marriage to Anne Cocke in 1732.[6] Together the couple had five sons, including William Woodford, their eldest child.[7]

Little is known about William Woodford's early years. It is likely, but uncertain, that he received a formal education, perhaps through tutors or perhaps through a more structured school setting. The lack of documentary evidence provides little guidance on the matter.

As the eldest son of a prominent landowner, merchant, and public servant in Caroline County, young William Woodford likely basked in and benefitted from his father's high social status. In 1755, however, William Woodford, just 21 years old, unexpectedly found himself thrust to the head of his family when his father suddenly passed away.

With a grieving mother and four younger brothers to tend to, one might expect Woodford to have remained at Windsor, but within months of his father's death he left his home and family to command a company of Caroline County militia in the French and Indian War.

---

[6] Catesby Willis Stewart, ed., *Woodford Letter Book: 1723-1737*, (Verona, VA: McClure Printing Company, Inc., 1977), 16

[7] Stewart, *The Life of Brigadier General William Woodford of the American Revolution*, Vol. 1, 70

French and Indian War

Conflict with France and her Indian allies on the colonial frontier was an ever present danger in the early 1750's, one that erupted into open warfare in 1754 when a detachment of Virginian militia commanded by 23 year old Lieutenant-Colonel George Washington (supported by a company of British regulars) fought and ultimately surrendered to a large French and Indian force in the Pennsylvania wilderness. This humiliating defeat at Fort Necessity, followed a year later by the defeat of General Edward Braddock's British army (which had marched into the Pennsylvania wilderness to, in part, avenge Washington's loss at Fort Necessity) triggered a global conflict between Britain and France and their allies. Virginians like William Woodford stepped forward to join the fight and in the spring of 1756 he reported to Colonel Washington at Fort Loudoun in Winchester at the head of a company of Caroline County militia.[8]

Captain Woodford and his 50 man militia company were posted at the confluence of Sleepy Creek and the Potomac River, about 45 miles north of Winchester.[9] They were to, *"protect the people from the insults of the enemy,"* and build earthworks to protect themselves from attack.[10] Washington instructed Woodford to send out frequent scouting parties to search for signs of the enemy, while at the same time discourage marauding and desertion among his own ranks by

---

[8] W.W.Abbot, ed., "Orders for the Militia, 15 May, 1756," *The Papers of George Washington, Colonial Series*, Vol. 3, (Charlottesville: University Press of Virginia, 1984), 137

[9] Ibid.

[10] Abbot, ed., "Colonel Washington to William Woodford, 16 May, 1756," *The Papers of George Washington, Colonial Series*, Vol. 3, 137

calling the company roll at least twice a day.[11] Unfortunately for Captain Woodford, such efforts did not prevent his men from deserting, and within six weeks all but eleven men had done so.[12]

Woodford's military service over the next year is unclear. With the bulk of his militia company gone it is possible that Woodford returned to Caroline County, but no account of his return exists. He very well may have remained in Winchester as a gentleman volunteer in Colonel Washington's Virginia Regiment, a provincial unit of colonial regulars. Woodford was referred to as such by Colonel Washington a year later in the summer of 1757.[13] Unlike the militia, who served for short periods of time, the troops of Colonel Washington's Virginia Regiment enlisted for a full year. They were Virginia's equivalent to British regulars.

By the middle of July of 1757, Colonel Washington had elevated Woodford from gentleman volunteer to ensign in the 1st Virginia Regiment.[14] He was attached to Captain Thomas Waggener's company, which was posted on the south branch of the Potomac River about 75 miles west of Winchester.[15] Waggener's orders were to protect the inhabitants of the

---

[11] Ibid.
[12] Abbot, ed., "Robert Stewart to Colonel Washington 3 July, 1756," *The Papers of George Washington, Colonial Series*, Vol. 3, 235-236
[13] W.W. Abbot, ed., "Colonel Washington to Robert Dinwiddie, 11 July, 1757," *The Papers of George Washington, Colonial Series*, Vol. 4, (Charlottesville: University Press of Virginia, 1984), 296
[14] Ibid.
[15] Abbot, ed., "Colonel Washington to Thomas Waggener, 29 July, 1757," *The Papers of George Washington, Colonial Series*, Vol. 4, 348-349

region from the Indians who, within a month of Waggener's arrival, staged several ineffectual raids.[16]

After a tense winter on the frontier, Woodford found himself embroiled in the spring of 1758 in a scandal involving one of his fellow officers. Lieutenant Peter Steenbergen had served in the Virginia Regiment since 1755. He was second in command of Captain Waggener's company and thus Woodford's superior officer. Steenbergen commanded a small detachment of Waggener's company in one of the smaller forts along the south branch of the Potomac.

In late April of 1758 Ensign Woodford, along with Captain Robert McKenzie, whose company was posted near Captain Waggener's, declared that they would no longer rank with Lieutenant Steenbergen due to conduct that they considered unbecoming an officer.[17] Woodford accused Steenbergen at a court of inquiry in early May of several attempts to profit at the expense of his men. Steenbergen denied the charges, but the court sided with Woodford.

Ensign Woodford assumed command of Steenbergen's detachment at a small outpost dubbed Fort George and discovered more evidence of Steenbergen's misconduct. He shared his discovery with Colonel Washington:

> *I have made it my Business since I Came here, to find out the past Conduct of Lieut Steenbergen. I can plainly make it Appear by his own Books which by*

---

[16] Abbot, ed., "Instructions to Company Captains, 29, July, 1757," *The Papers of George Washington, Colonial Series*, Vol. 4, 341-345 and "Colonel Washington to Robert Dinwiddie, 27 August, 1757," 385

[17] W.W. Abbot, ed., "Court of Inquiry 4-8 May, 1758," *The Papers of George Washington, Colonial Series*, Vol. 5, (Charlottesville: University Press of Virginia, 1988), 162-163

good fortune have fallen into my Hand, that he has Defrauded the Soldiers of their Pay in the most scandalous Manner. As this is more then he would declare, when Examined at Winchester, I thought it my Duty...to Acquaint you with it that you might take the proper measures to bring him to Justice.[18]

It is unknown how Colonel Washington responded to Woodford's report; more pressing issues overshadowed Steenbergen's alleged misconduct in the summer of 1758.

Four years had passed since Colonel Washington had tried to secure Virginia's claims to western Pennsylvania by establishing a fort at the point of land where the Monongahela and Alleghany Rivers converged to form the Ohio River (present day Pittsburg). A French fort (Fort Duquesne) stood at this strategic location instead, and British leaders were determined to capture it.

In preparation for this, Colonel Washington recalled the scattered companies of the 1st Virginia Regiment and assembled them at Fort Cumberland, in western Maryland in July, 1758.[19] They were to join a British expedition commanded by General John Forbes and help build and secure a new road through the Pennsylvania wilderness (protected by a chain of forts) straight to Fort Duquesne.

---

[18] Abbot, ed., "William Woodford to Colonel Washington, 17 May, 1758," *The Papers of George Washington, Colonial Series*, Vol. 5, 187
[19] Abbot, ed., "Return of Tents, 28 July, 1758," *The Papers of George Washington, Colonial Series*, Vol. 5, 348

Newly promoted Lieutenant William Woodford participated in this expedition as an officer in Captain Robert Stewart's company of 1st Virginians.[20] Captain Stewart was the senior captain of Washington's regiment and a Scotsman, highly esteemed by his fellow officers and the British officer corps.

General Forbes's troops made slow but steady progress towards Fort Duquesne in the fall of 1758, suffering just one significant setback in September with an ill-advised attack on Fort Duquesne by British Major James Grant and 800 Scottish and colonial advance troops who had arrived at the fort well ahead of General Forbes and the main body of troops. Grant's rash attack was easily repulsed by the French and their Indian allies. Undeterred by this defeat, General Forbes slowly but steadily marched westward with his main body. The approach of winter caused him to halt in mid-December, well to the east of Fort Duquesne. When he learned just days later, however, that British negotiators had signed a treaty with many of the Indian nations allied with the French (causing them to abandon the French) Forbes disregarded the threat of winter and pushed westward with renewed vigor towards Fort Duquesne.

Without their Indian allies, the French could no longer hold the fort, so they abandoned it, setting it ablaze when they left. British and colonial troops took possession of the ruined fort just hours after it was abandoned.

---

[20] W.W. Abbot, ed., "Orderly Book, 1 November, 1758," *The Papers of George Washington, Colonial Series*, Vol. 6, (Charlottesville: University Press of Virginia, 1988), 102

Within a month of the fall of Fort Duquesne (renamed Fort Pitt in honor of the British Prime Minister, William Pitt) Lieutenant Woodford and most of the Virginia Regiment were back at Fort Loudoun in Winchester. Soon after their arrival they received upsetting news. Colonel Washington, who had departed for Williamsburg, had decided to resign his commission and leave the regiment. Lieutenant Woodford joined his fellow officers in a written address to Washington that heaped praise and best wishes on their departing commander.[21]

Unfortunately, the level of documentary evidence regarding Lieutenant Woodford's activities with the Virginia Regiment diminished dramatically with Colonel Washington's departure from the war. No other Virginian's involvement and interactions in the conflict are as well documented as Washington's.

It appears though, that William Woodford remained in the service of the 1st Virginia Regiment as a lieutenant for the next three years. After a very difficult winter in Winchester (in which the regiment endured severe shortages of food medicine, clothing, and pay) the Virginia Regiment, under the command of Colonel William Byrd, marched to Fort Ligonier in Pennsylvania.[22] Their stay there was brief, for by June

---

[21] Abbot, ed., "Address from the Officers of the Virginia Regiment, 31 December, 1758," *The Papers of George Washington, Colonial Series*, Vol. 6, 178-180

[22] George Reese, ed., "William Byrd to Governor Francis Fauquier, 22 March, 1759," *The Official Papers of Francis Fauquier, Lt. Governor of Virginia, 1758-1760*, Vol. 1, (Charlottesville: University Press of Virginia, 1980), 192

1759 the regiment, some 800 strong, was back in Winchester.[23]

The fortunes of war continued to favor Britain and her colonies in 1759. They captured two critical French posts, Fort Niagara in July and Quebec in September. The loss of these posts significantly weakened the French hold on Canada and prompted even more of their Indian allies to abandon them.

Lieutenant Woodford and the Virginia Regiment returned to Pennsylvania in July to help defend Fort Bedford and Fort Ligonier.[24] Several small skirmishes with Indian parties occurred, but the bulk of the remaining hostile Indians in the area retreated westward to Detroit after the French surrender of Fort Niagara. As a result, relative peace settled over the region.

Woodford remained with the Virginia Regiment, which returned to Virginia, for two and a half more years. With the French largely defeated it seemed there was little to threaten the colonists. The House of Burgesses, eager to reduce expenses, reluctantly maintained the regiment (largely as a deterrent to the Cherokee in the southwestern region of the colony) but a treaty with the Cherokee in 1761 eliminated the need for such troops. Lieutenant Woodford's last assignment with the Virginia Regiment was to escort a party of Cherokee leaders to Williamsburg to finalize the treaty. He left Fort Chiswell in southwestern Virginia with the Indian leaders in

---

[23] Reese, ed., "Governor Fauquier to the Board of Trade, 9 June, 1759," *The Official Papers of Francis Fauquier, Lt. Governor of Virginia, 1758-1760*, Vol. 1, 218-219

[24] Reese, ed., "Governor Fauquier to the Board of Trade, 2 August, 1759," *The Official Papers of Francis Fauquier, Lt. Governor of Virginia, 1758-1760*, Vol. 1, 230

late December 1761 and reached Williamsburg the following month.[25] Upon his arrival Lieutenant Woodford learned that the House of Burgesses had voted to disband the regiment in the spring, so instead of returning westward, Woodford headed home to Caroline County.

Woodford Returns to Windsor

Six years had passed since William Woodford first left Windsor and during that time he had developed into an experienced military officer. With his service concluded, the twenty-seven year old patriarch of Windsor returned to the comfortable life of a Virginia planter.

Within a few months of his return Woodford married eighteen year old Mary Thornton, the daughter of his neighbor, Colonel John Thornton of Snow Creek and a distant cousin of George Washington.[26] The couple's first son, John Thornton Woodford, was born a year after their marriage. A daughter followed two years later, but died within a week of her birth. A year later, in 1766, a second son, William Catesby, completed Woodford's family.[27]

William Woodford focused on more than just personal matters after his return from the French and Indian War. He wished to follow his father's example of public service and was elected to the vestry of St. Mary's parish in 1763. Two years later, Governor Francis Fauquier named Woodford a

---

[25] Reese, ed., "Lt. Col. Adam Stephen to Gov. Francis Fauquier, 22 December, 1761," *The Official Papers of Francis Fauquier, Lt. Governor of Virginia, 1758-1760*, Vol. 2, 626
[26] Stewart, *The Life of Brigadier General William Woodford of the American Revolution*, Vol. 1, 17, 218-219
[27] Ibid., 232-234

justice of the peace for Caroline County.[28] In these positions Woodford participated in the management of the local church parish, which included providing aid to the poor and disabled, and also enforced the law as a justice of the peace.

New British Policies Provoke Colonial Resistance

Family and local affairs were not the only concerns of William Woodford in the 1760's. Disturbing reports of new British policies aimed at the American colonies (policies many colonists viewed as unconstitutional) began to appear in the weekly gazettes in the mid-1760's. The 1765 Stamp Act was the most notorious of these new policies and ignited a firestorm of protest throughout Virginia.

Virginians were long accustomed to paying taxes levied by their own representatives in the House of Burgesses and church vestries (internal taxes) to assist the poor and pay for the operation of government and the debt accrued from the French and Indian War.[29] As inconvenient and burdensome as such taxes and church levies were they were accepted by Virginians as legitimate because they were authorized by representatives selected by Virginians.

The 1765 Stamp Act, however, was not viewed in the same light. It was a tax imposed on the colonists by the British Parliament, a body that most Virginians believed in no way represented them. If the stamp tax was allowed to stand it would establish a dangerous precedent, taxation of the

---

[28] Ruth and Sam Sparacio, eds., " Caroline Court, 13 June, 1765," *Virginia County Court Records Order Book, Caroline County, Virginia, 1765,* (The Antient Press, 1989), 1

[29] Reese, ed., "Report on the Colony, 30 January, 1763," *The Official Papers of Francis Fauquier*, Vol. 2, 1009-1022

colonists by an entity that did not represent them. This would signal a dangerous expansion of parliamentary power over the colonies, power that the colonists would be unable to resist. The outcry from the colonists against the Stamp Act was thus loud, adamant, and universal.

The freeholders of Westmoreland County, just across the Rappahannock River on the Northern Neck of Virginia, expressed the fear of most Virginians when they declared that the stamp tax was an attempt to *"reduce the People of this Country to a State of abject and detestable slavery."*[30] Colonel George Washington of Fairfax County stressed the unconstitutional elements of the act and stated that Virginians

> *Look upon this unconstitutional method of Taxation as a direful attack upon their Liberties, and loudly exclaim against the Violation.*[31]

It is uncertain how William Woodford reacted to the Stamp Act; it is probable that his reaction was similar to most of his friends and associates, namely, he strongly opposed it. He also likely applauded the news in 1766 of Parliament's repeal of the Stamp Act. It appeared that the colonists' concerns had been addressed by Parliament, but only temporarily.

Desperate for a way to shift some of the financial costs of the empire to the colonies and determined to re-assert its

---

[30] William Van Schreeven and Robert L. Scribner, eds. "Resolutions of the Westmoreland Association in Defiance of the Stamp Act, 27 February, 1766," *Revolutionary Virginia: The Road to Independence,* Vol. 1, (University Press of Virginia, 1973), 24

[31] W.W. Abbot and Dorothy Twohig, eds. "George Washington to Francis Dandridge, 20 September, 1765," *The Papers of George Washington: Colonial Series,* Vol. 7, (Charlottesville: University Press of Virginia, 1990), 395-396

authority over the colonists, Parliament adopted the Townshend Duties in 1767. This act placed tariffs on a list of goods shipped to the colonies from Britain. The newly taxed items included tea, glass, paint, and paper. The British Ministry argued that since the colonists had long accepted Parliament's right to tax imports to the colonies, they could not legally oppose the new duties because the goods being taxed were imported from Britain to the colonies. Concerned that these new measures might lead to a repeat of the unrest that occurred in Boston by opponents of the Stamp Act, the British Ministry sent several regiments of troops to Boston to keep order.

The military occupation of Boston in 1768 helped spur Virginians like Colonel George Washington and William Woodford to support stronger opposition to Parliament's policies. In the spring of 1769, Colonel Washington proposed a non-importation plan drafted by his friend, George Mason of Fairfax County. Mason believed that a boycott would cause English manufacturers to pressure Parliament to redress American grievances.

When Virginia's governor, Lord Botetourt, learned of Mason's plan he dissolved the House of Burgesses to prevent its consideration. Undeterred, the dismissed burgesses met at a nearby tavern and approved most of Mason's proposals. Ninety-four burgesses joined a voluntary association to boycott the taxed items.[32] Thousands of Virginians, including William Woodford, followed their example and subscribed to

---

[32] Van Schreeven and Scribner, eds., "Non-importation Resolutions of the Former Burgesses, 18 May, 1769," *Revolutionary Virginia: The Road to Independence,* Vol. 1, 76-77

the non-importation association of 1769.³³ Most of the other colonies adopted similar measures, and by the fall of 1769, a continent wide boycott on select British goods was implemented. Some Virginians were reluctant to sacrifice their English comforts. To counter this, county committees were authorized to search for contraband (boycotted goods). They were also instructed to publish the names of boycott violators in the Virginia gazettes.

William Woodford was one of five men selected among nearly 350 subscribers to the non-importation association in Caroline County to serve on a committee of inspection to enforce the association.³⁴ The need for this committee, however, disappeared soon after it was formed when news of Parliament's repeal of the Townshend Duties reached Virginia in the summer of 1770. Parliament's abandonment of the Townshend Duties was prompted more by economic considerations than political ones; the duties had failed to raise much revenue for the treasury. Despite the repeal of the duties, the British Ministry and its supporters continued to claim that Parliament had the authority to tax the colonists. To emphasize this point the tax on tea was maintained.

Boston Tea Party 1773

The tea tax became the focal point of renewed tensions between Britain and her American colonies in 1773 when Parliament attempted to help the East India Company unload a surplus supply of tea that had accumulated in England. On the

---

³³ William Rind, *Virginia Gazette*, 17 January, 1771, 2
³⁴ Ibid.

surface the Tea Act of 1773 appeared to benefit the colonists. East India Company agents in the colonies were given a monopoly to sell taxed tea directly to the colonists at a lower price than all of their competitors, even those who sold smuggled Dutch tea.

The potential financial savings of this new arrangement for the colonists created a dilemma for colonial leaders. They feared that Parliament would interpret a spike in tea sales as tacit colonial acceptance of the tea tax. This would be the precedent Parliament needed to reaffirm its right to tax the colonists. Colonial leaders were determined to prevent this from happening and undertook efforts to block the delivery of the surplus tea in Charleston, Philadelphia, New York, and Boston. They succeeded without mishap in all the ports except Boston. On December 16th, 1773, scores of Bostonians took the extreme measure of dumping a large shipment of imported tea into Boston Harbor. This action infuriated Parliament and the King; they interpreted the act as a blatant challenge to their authority and responded forcefully.

Outraged by the "Boston Tea Party", the British Parliament moved to punish the entire city of Boston with passage of the Boston Port Act. This act, which was one in a series of measures dubbed the Intolerable Acts by the colonists, closed Boston harbor to all trade and commerce. British warships and soldiers were sent to Boston to enforce the port closure and help keep order in the "lawless" city. The port would only be reopened upon payment for the destroyed tea.

Virginians React to the Intolerable Acts 1774

As word spread about Parliament's harsh response to the "tea party" in Boston, more and more Virginians concluded that the British measures were a dangerous abuse of parliamentary power. Edmund Pendleton, a moderate burgess from Caroline County, expressed the view of many Virginians (including in all probability his close friend William Woodford) when he declared that

> *Tho' it should be granted that the Bostonians did wrong in destroying the tea, yet the Parliament giving Judgement and sending ships and troops to* [punish the entire city] *in a case of Private property is* [an] *Attack upon constitutional Rights, of which we could not remain Idle Spectators.....*[35]

The House of Burgesses initially demonstrated its support for Boston by calling for a day of public prayer. The Royal Governor of Virginia, John Murray, the Earl of Dunmore, viewed this action as an insult to the King and Parliament and dissolved the burgesses before the assembly had finished its legislative business. Many of the dismissed and defiant burgesses gathered at a tavern near the capitol and agreed that a general meeting of colonial representatives was needed to develop an effective united response to Parliament.[36]

---

[35] David John Mays, ed., "Edmund Pendleton to Joseph Chew, 20 June, 1774," *The Letters and Papers of Edmund Pendleton*, Vol. 1 (Charlottesville: University Press of Virginia, 1967), 93

[36] Van Schreeven and Scribner, eds.,"An Association Signed by 89 Members of the late House of Burgesses, 27 May 1774," *Revolutionary Virginia: The Road to Independence*, Vol. 1, 97-98

Colonial leaders from New England and the mid-Atlantic colonies had the same idea and a general congress was scheduled in Philadelphia in September to discuss the crisis. Each county in Virginia was instructed by the former Speaker of the House of Burgesses (Peyton Randolph) to send two representatives to Williamsburg in early August to attend a special convention that would select and instruct Virginia's delegates to the Continental Congress.

Freeholders throughout Virginia met at their county courthouses during the summer to adopt resolutions of support for Boston and to select delegates to the Convention. William Woodford, who owned property in Fredericksburg as well as in Caroline County, was chosen to serve on Fredericksburg's committee of correspondence on June 1st, 1774. Six weeks later he participated in a general meeting of freeholders of Caroline County that adopted resolutions in support of Massachusetts.[37]

The Caroline County resolves which Captain Woodford likely helped write, began with a pledge to preserve, *"all due Obedience and Fidelity,"* to the King and his government and maintain, *"a firm union,"* with Great Britain.[38] Woodford and his fellow freeholders then turned to their grievances with Parliament. They asserted that taxation without representation violated both their constitutional and natural rights as Englishmen and that Parliament had been influenced by, *"evil Counsellors"* who had departed from the true principles of the British Constitution.[39] They urged unity among all the

---

[37] Rind, *Virginia Gazette*, 16 June, 1774, 2, and Alexander Purdie and John Dixon, *Virginia Gazette*, Supplement, 28 July, 1774, 1
[38] Purdie and Dixon, *Virginia Gazette*, Supplement, 28 July, 1774, 1
[39] Ibid.

colonies and endorsed both a meeting of a continental Congress in Philadelphia and a general boycott of British goods in support of Massachusetts. They also called for a ban on the African slave trade, which was deemed, "*Injurious to this Colony*," by suppressing the emigration of freemen from Europe.[40] The meeting concluded with the selection of Edmund Pendleton and James Taylor to represent Caroline County at the Virginia Convention in Williamsburg in August.[41]

When the 1st Virginia Convention delegates assembled in Williamsburg three weeks later, they selected Peyton Randolph, George Washington, Patrick Henry, Richard Henry Lee, Edmund Pendleton, Benjamin Harrison, and Richard Bland to represent Virginia in the general Congress in Philadelphia and sent instructions with the delegates that largely mirrored the resolves passed by Caroline County three weeks earlier.

First Continental Congress

The First Continental Congress convened in Philadelphia on September 5th, and chose Peyton Randolph to preside over it. After several procedural matters were settled, Congress tackled the main issue, a unified colonial response to the Intolerable Acts.

Some of the more militant delegates, such as John Adams, Patrick Henry, and Richard Henry Lee, proposed that the Congress urge each colony to better organize and discipline its militia. Opponents of this idea called the proposal a

---

[40] Ibid.
[41] Ibid.

declaration of war on the British Ministry that undermined the true purpose of the Continental Congress, which was to adopt the best means to affect a reconciliation with Britain.[42] They preferred trade restrictions on both imports and exports that would impact British manufacturers and spur them to pressure Parliament to resolve the dispute peacefully. After several days of debate, the militia proposals were defeated and, much to the chagrin of Adams, Henry, Lee, and a few other delegates, Congress shifted its attention solely to economic measures against Britain.

Two more weeks passed before the delegates finally agreed on a specific course of action. The plan they adopted in late October was dubbed the Continental Association and was similar to the proposals from Caroline County and the Virginia Convention earlier that summer. A boycott of British goods and the discontinuation of the slave trade were approved for December 1st, 1774. If Parliament persisted with the Intolerable Acts, a ban on all colonial exports to Britain would follow on September 1st, 1775. Congress also called on the American colonists to be more frugal and industrious and to avoid extravagant activities like horse-racing, gambling, plays, and dances. Committees were authorized in every county, city, and town to enforce these provisions.[43]

Parliament's plan had backfired. Instead of intimidating the colonists with its harsh treatment of Massachusetts, the

---

[42] Paul H. Smith, "Silas Deane's Diary, 3 October, 1774," *The Letters of Delegates to Congress*, Vol. 1, (Library of Congress, 1976), 138

[43] Worthington C. Ford, "20 October, 1774," *Journals of the Continental Congress: 1774-1789,* Vol. 1, (Washington: Government Printing Office, 1904), 75-80

American colonists rallied to Boston's defense and became more united.

Chapter Two

"I Highly Approve of Your Appointment"

1774-1775

Caroline County was one of the first counties in Virginia to act on the Continental Congress's resolutions. On November 10th, 1774, the freeholders of Caroline County selected 20 gentlemen, including William Woodford and Edmund Pendleton, to serve on a committee to enforce the Continental Association.[1] A few Virginia counties north of Caroline went further in their opposition to Britain.

Independent Militia Companies

Laws for the regulation of the militia in Virginia dated back to the 1730's and were periodically renewed by the House of Burgesses (most recently in 1771) but when Governor Dunmore dissolved the House of Burgesses in May 1774 he denied the assembly the chance to renew the recently expired militia law. This meant that the authority to assemble, discipline and train the various county militias no longer existed in Virginia and as a result, Virginia's militia was in some disarray.

---

[1] H.R. McIlwaine, ed., "Proceedings of the Committees of Safety of Caroline and Southampton Counties, Virginia: 1774-1776, *Bulletin of the Virginia State Library*, Vol. 17, No. 3, (November, 1929), 127

Although the Continental Congress refused to encourage the strengthening of colonial militia, George Mason was determined to do so for Fairfax County. In September, while the delegates in Philadelphia debated the militia proposal of John Adams and Richard Henry Lee, Mason persuaded the freeholders of Fairfax County to form an independent militia company. Membership in the company was open to the first one hundred men who met the unit's strict requirements. Colonel George Washington accepted command of the company upon his return from Congress and those who joined it agreed to outfit and equip themselves at their own expense and meet regularly to drill.[2]

Fairfax County was not the only county in the fall of 1774 to take such a militant stand. Neighboring Prince William County also formed an independent militia company in November and by December, two additional counties, Loudoun and Spotsylvania, moved to raise their own independent companies.[3]

---

[2] Robert A. Rutland, ed., "Fairfax County Militia Association 21 September, 1774," *The Papers of George Mason*, Vol. 1, (University of North Carolina Press, 1970), 210-211

[3] Stanislaus M. Hamilton, ed. "Extract from the Minutes of the Independent Company of Cadets of the 11th, November, 1774," *Letters to Washington & Accompanying Papers*, Vol. 5 (Boston & New York: Houghton Mifflin, Co., 1902, 68-69
and
   Nicholas Cresswell, "13 December, 1774," *The Journal of Nicholas Cresswell,* (New York: The Dial Press, 1974), 51
and
   William J. Van Schreeven and Robert L. Scribner, eds., "Spotsylvania County Committee, 15 December, 1774," *Revolutionary Virginia: The Road to Independence,* Vol. 2, (University Press of Virginia, 1975), 196-197

While several of the counties north of Caroline formed independent militia companies, William Woodford and the freeholders of Caroline focused primarily on enforcement of the Continental Association. In mid-December, Woodford was appointed to a committee of inspection (comprised of eight gentlemen) charged with inspecting the books of merchants in Port Royal to insure that they had not raised their prices in response to the Continental Association.[4]

Caroline County Independent Militia Company

Reports in the weekly gazettes in December and January of a British ban on gunpowder and arms shipments to the colonies may have prompted the Caroline Committee to form its own independent militia company in January 1775. In a meeting of the county committee that month, William Woodford was listed in the minutes as Captain Woodford and was named as one of three committee members with the authority to call a meeting of the committee. The sudden reference to Captain Woodford suggests that the committee may have formed an independent militia company in January and appointed Woodford (one of the most militarily experienced gentlemen in Caroline County) to lead it.[5]

If so, Captain Woodford, despite his new responsibilities as a militia commander, continued to serve on the county committee. In February, Woodford and four other committee members were authorized to sell coal that had recently arrived from Britain (in violation of the Continental Association) and use the proceeds to purchase gunpowder. Woodford and his

---

[4] McIlwaine, ed., "Caroline County Proceedings, 16 December, 1774," 127

[5] McIlwaine, ed., "Caroline County Proceedings, 19 January 1775" 129

fellow committee members were also instructed to, "*collect from every person*," funds to purchase more powder.[6] A month later, Captain Woodford and James Taylor purchased five barrels of powder on behalf of the Caroline Committee.[7] Clearly, Caroline County, like much of Virginia, was preparing for an armed conflict.

2nd Virginia Convention

In late March, delegates from nearly all of Virginia's counties and towns assembled in Richmond to select representatives for a second Continental Congress in Philadelphia. On the fourth day of the meeting Patrick Henry, the firebrand from Hanover County, seized the opportunity to propose that Virginia *"Be immediately put into a posture of defense."*[8] Henry believed more than ever that armed conflict with Britain was inevitable and he urged Virginia to assume a war footing.

Henry's proposal to expand the colony's military preparedness was opposed by moderates like Robert Carter Nicholas and Edmund Pendleton (William Woodford's close friend). Nicholas, Pendleton, and other opponents argued that Henry's proposal was too confrontational and would antagonize Parliament and perhaps provoke a war. They urged patience and scoffed at the idea of fighting Britain.[9]

---

[6] McIlwaine, ed., "Caroline County Proceedings, 9 February 1775" 130
  Note: The committee voted on 13 April, 1775 to collect two shillings for every Tithe in the County.
[7] McIlwaine, ed., "Caroline County Proceedings, 8 March 1775" 130
[8] Van Schreeven and Scribner, eds., "Proceedings of the Second Virginia Convention, 23 March, 1775," *Revolutionary Virginia: The Road to Independence,* Vol. 2, 366-367
[9] William Wirt, *Sketches in the Life and Character of Patrick Henry,* (Philadelphia, 1817), 136

Patrick Henry replied to his critics and in doing so delivered one of the greatest speeches in American history. He acknowledged the patriotism of the opposition but suggested that they had, "*shut* [their] *eyes against a painful truth.*"[10] Henry was alarmed by Britain's huge military buildup in Massachusetts and asked the delegates whether, "*fleets and armies* [were] *necessary...* [for] *reconciliation.*"[11] Henry recounted the failure of a decade's worth of pleas and petitions to Britain and asserted that conflict with Great Britain was inevitable:

> *Gentlemen may cry, peace, peace – but there is no peace. The war is actually begun! The next gale that sweeps from the north will bring to our ears the clash of resounding arms! Our brethren are already in the field! Why stand we here idle? What is it that gentlemen wish? What would they have? Is life so dear, or peace so sweet, as to be purchased at the price of chains and slavery? Forbid it, Almighty God! I know not what course others may take; but as for me, give me liberty, or give me death!*[12]

Henry's stirring appeal to expand the militia worked and his resolution passed by a narrow margin.

---

[10] Ibid. 138
[11] Ibid. 139
[12] Ibid.

The Gunpowder Incident

Just a month later, in the early morning hours of April 21st, 1775, a detachment of British Marines and sailors crept into Williamsburg, removed fifteen barrels of gunpowder from the powder magazine in the center of the capital, and delivered them to the *H.M.S. Magdalen* anchored in the James River. Governor Dunmore, concerned by the actions of the 2nd Virginia Convention in Richmond a month earlier had arranged for the seizure of the powder.

Residents of Williamsburg were outraged at this action and gathered at the courthouse, which was across from the powder magazine, to demand action. Speaker Randolph and a handful of other local leaders dissuaded the crowd from marching on the Governor's Palace to threaten Dunmore and instead met with Dunmore to request the return of the powder. Dunmore gave a feeble explanation for his actions and promised to deliver the powder as soon as it was needed, but he refused to return it to the magazine, which he claimed was not secure.

Surprisingly, the large crowd that had gathered at the courthouse to await Dunmore's reply acquiesced to Speaker Randolph's pleas for calm and dispersed without further incident. The same was not true outside of Williamsburg, where the news of Dunmore's actions spread quickly, punctuated by reports of his threat to burn down the capital and to arm his slaves if a mob dared threaten Dunmore or his family again.

News of the seized powder alarmed the Virginia countryside. Hundreds of militia from multiple counties gathered in Fredericksburg with the intention to march on the capital to confront Lord Dunmore and demand the gunpowder's return. It is very likely that Captain Woodford was in Fredericksburg at this time on behalf of the Caroline County militia company, his men waiting anxiously in Bowling Green for word on whether they were to march to Williamsburg or not.

Dispatch riders sent from Fredericksburg to Williamsburg to obtain more information returned to Fredericksburg on April 28th, and reported that Peyton Randolph and the city leaders of Williamsburg had the situation under control. The messengers added that Speaker Randolph, concerned with the volatile temper of Governor Dunmore, urged the militia in Fredericksburg to abandon their planned march on the capital.[13] Randolph feared it would only provoke Dunmore and exacerbate the crisis.

An officer's council (representing 600 mounted militia) listened intently to Speaker Randolph's message, then engaged in a heated debate on how to proceed. As Peyton Randolph was the most respected man in Virginia, his views held tremendous weight with the officers and after much discussion it was agreed to heed Randolph's plea and cancel the march to the capital. Instead, the officer's council publically condemned Lord Dunmore for his actions and pledged to re-assemble the militia at a moment's notice to defend their rights or those of any sister colony that was unjustly invaded.[14]

---

[13] Robert L. Scribner, ed., *Revolutionary Virginia: The Road to Independence* Vol. 3, (University Press of Virginia, 1977), 63-64
[14] Scribner, ed., "Spotsylvania Council, 29 April, 1775," *Revolutionary Virginia: The Road to Independence*, Vol. 3, 71

On the same day that the militia in Fredericksburg cancelled their march (April 29th) the Caroline County Committee met and recommended that Captain Woodford's company of militia remain in Caroline County until Speaker Randolph (who was due to pass through Caroline County on his way to the 2nd Continental Congress in Philadelphia) could be consulted.[15]

While Captain Woodford and his troops awaited the arrival of Speaker Randolph, Patrick Henry went ahead with his own march on Williamsburg, leading 150 militia from Hanover, New Kent, and King William counties towards Williamsburg in early May. Henry halted only a day's march from the city. Two days of negotiations ensued with intermediaries attempting to defuse the crisis before Henry agreed to accept payment for the seized gunpowder in lieu of its return. Henry dismissed the militia and returned home to prepare for his trip to Philadelphia to attend the Congress.

Caroline County Militia and Committee

Captain Woodford and his company of militia, still in Caroline County, performed one more duty before they dispersed and returned to their homes. They escorted Speaker Randolph, Benjamin Harrison, and Edmund Pendleton (who were delegates to the 2nd Continental Congress) northward to the Potomac River.[16]

Pendleton demonstrated his appreciation and support of Captain Woodford's company a few weeks later by procuring a stand of colors, two fifes, and a drum for the company in

---

[15] McIlwaine, ed., "Caroline County Proceedings, 29 April 1775" 131
[16] Purdie, *Virginia Gazette*, 12 May, 1775, Supplement, 1-2

Philadelphia. Pendleton also obtained a large officer's marquee tent on behalf of Captain Woodford.[17] Back in Caroline County, Captain Woodford resumed his work on the county committee. It met on May 9th, and publically thanked Patrick Henry for his efforts concerning the gunpowder incident.[18] Two days later the committee selected Woodford to serve as Edmund Pendleton's alternate in the next convention, scheduled for July in Richmond.[19] A few weeks later the Caroline Committee authorized Woodford to purchase up to 1000 pounds of gunpowder and pay for it with the money collected throughout the county by the committee.[20]

Dunmore Flees

On the same day that the Caroline committee authorized the gunpowder purchase, a new crisis developed in Williamsburg. May had been a tense month for Governor Dunmore and his family in the capital. The seizure of the gunpowder in late April had ignited several confrontations and news from the north of bloodshed in Massachusetts (at Lexington and Concord) as well as militia marching on Williamsburg only heightened tensions. In early June the governor concluded that it was best to flee the capital with his family and he did so on the evening of June 8th. They found refuge on a British warship anchored in the York River.

---

[17] David John Mays, ed., "Edmund Pendleton to William Woodford, 30 May, 1775," *The Letters and Papers of Edmund Pendleton, 1734 – 1803,* Vol. 1, 103
[18] Purdie, *Virginia Gazette*, 19 May, 1775, Supplement, 3
[19] McIlwaine, ed., "Caroline County Proceedings, 11 May 1775" 131
[20] McIlwaine, ed., "Caroline County Proceedings, 8 June 1775" 131

Weeks of fruitless appeals by the House of Burgesses for Dunmore to return ensued, but the governor refused, essentially ending royal governance in the colony. Unable to complete any legislation, the burgesses adjourned in late June, never to meet as the House of Burgesses again.

Just three weeks later, in mid-July, Virginia's third convention of county representatives assembled in Richmond and effectively served as the new governing authority in Virginia. Among the delegates in attendance was Captain William Woodford of Caroline County.

Delegate to the 3rd Virginia Convention

Unlike the first two conventions held in Virginia in August of 1774 and March of 1775, the 3rd Virginia Convention was not convened primarily to select and instruct delegates to the next Continental Congress. This convention devoted the bulk of its efforts to strengthening Virginia's defenses. Captain Woodford's extensive military service in the French and Indian War secured him a spot on a committee formed to, *"Oversee that a sufficient Force be immediately raised and embodied...for the Defense and protection of the colony."* Woodford played an important role in the committee's efforts.[21]

It appears that during committee deliberations Captain Woodford's military experience and knowledge attracted the notice of many members, so much so that on August 3rd the committee recommended to the convention that 1,000 regular (full time) troops be raised in Virginia under the command of

---

[21] Scribner, ed., "Proceedings of the Third Virginia Convention, 19 July, 1775," *Revolutionary Virginia: The Road to Independence*, Vol. 3, 319

Woodford.[22] George Washington would have undoubtedly been the convention's first choice to command Virginia's troops, but he was in Massachusetts in command of the continental army. Another committee proposal called for Woodford to command 500 regular troops in the vicinity of Norfolk and Portsmouth while Thomas Nelson Jr. of Yorktown assumed command of another 500 such troops in Williamsburg.[23] Supporters of Patrick Henry, who desired command of Virginia's forces himself despite his lack of military experience, successfully delayed votes on these proposals and they were eventually replaced by other proposals.

On August 5th, after much debate and despite the fact that no resolution had been reached on the number of troops to be raised, the 3rd Virginia Convention selected commanders for three regiments of regular troops. It took two ballots for the convention to choose the colonel of the 1st Virginia Regiment (and de facto commander-in chief of Virginia's regular forces). Hugh Mercer of Fredericksburg, a veteran of the French and Indian War, held a one vote lead over Patrick Henry, 41 to 40 with eight votes going to Thomas Nelson Jr. of Yorktown and a single vote to William Woodford.[24] The winner needed a majority of the votes, so those cast for Nelson and Woodford became the deciding votes. On the second ballot, six of the nine votes in question went to Henry, pushing

---

[22] Scribner, ed., "Proceedings of the Third Virginia Convention, 3 August, 1775," *Revolutionary Virginia: The Road to Independence*, Vol. 3, 393

[23] Scribner, ed., "Proceedings of the Third Virginia Convention, 3 August, 1775," Footnote 8, *Revolutionary Virginia: The Road to Independence*, Vol. 3, 395

[24] Scribner, ed., "Proceedings of the Third Virginia Convention, 5 August, 1775," *Revolutionary Virginia: The Road to Independence*, Vol. 3, 400

him over the top and making him colonel of the 1st Virginia Regiment.[25]

Command of the 2nd Virginia Regiment should presumably have gone to Mercer, especially given his extensive military experience as commander of Pennsylvanian troops in the late war with France. William Woodford, who knew Mercer well, actively supported Mercer's appointment to command the 2nd Regiment, declaring that he would gladly serve under Mercer, "*as he knew him to be a fine officer.*"[26] Mercer's Scottish heritage, however, suddenly became an issue to a number of delegates, many of who supported Mercer when there was a chance that he could stop the inexperienced Henry from gaining command, but who now had little use for Mercer and preferred a native Virginian in command of the 2nd Regiment over a "North Briton". So the Convention settled on Thomas Nelson Jr. for command of the 2nd Regiment.[27] A relatively close contest for command of a third regiment of regulars was won by William Woodford (over Patrick Henry's brother-in-law, William Christian) after two ballots.[28]

---

[25] Ibid.
[26] Scribner, ed., "Proceedings of the Third Virginia Convention, 5 August, 1775," Footnote 6, *Revolutionary Virginia: The Road to Independence*, Vol. 3, 403
[27] Scribner, ed., "Proceedings of the Third Virginia Convention, 5 August, 1775," *Revolutionary Virginia: The Road to Independence*, Vol. 3, 400 and Purdie, *Virginia Gazette,* 9 February, 1776, 3
[28] Scribner, ed., "Proceedings of the Third Virginia Convention, 5 August, 1775," *Revolutionary Virginia: The Road to Independence*, Vol. 3, 400-401

Several days later, on August 9th, Woodford relinquished his seat in the Convention to Edmund Pendleton, who arrived from Philadelphia.[29] That same day, the Convention essentially eliminated Woodford's regiment and command by reducing the number of regular troops to be raised from 3,000 to 1020.[30]

Undoubtedly disappointed, Woodford returned to Caroline County where the very next day he mustered and exercised the independent militia company. A correspondent to Alexander Purdie's gazette observed that Captain Woodford's troops performed the, *"manual exercise, with a great variety of new and useful evolutions, at the Bowling Green...before upwards of 1500 spectators, who were exceedingly pleased with the dexterity and alertness of the men...."*[31]

Woodford Gets Command of the 2nd Virginia Regiment

A week passed before Captain Woodford learned that Thomas Nelson Jr. had declined command of the 2nd Regiment (citing poor health) and that the Convention had appointed Woodford to command in Nelson's place.[32] Lieutenant Colonel Charles Scott of Cumberland County and Major

---

[29] Scribner, ed., "Proceedings of the Third Virginia Convention, 9 August, 1775," *Revolutionary Virginia: The Road to Independence*, Vol. 3, 409
[30] Ibid.,
[31] Purdie, *Virginia Gazette*, 11 August, 1775, 2
[32] Scribner, ed., "Proceedings of the Third Virginia Convention, 17 August, 1775," *Revolutionary Virginia: The Road to Independence*, Vol. 3, 457-458

Alexander Spotswood of Spotsylvania County (Woodford's close friend) joined Colonel Woodford in the 2nd Regiment.[33]

To recruit a sufficient number of regular (full time) troops, the 3rd Convention organized Virginia into sixteen districts, lumping counties together based on their size and proximity. Each district was instructed to recruit and send a company (sixty-eight men) of regular troops to Williamsburg as soon as possible to serve for a year.[34] The regular company raised on the eastern shore was ordered to remain in its district because the area remained vulnerable to attack from the sea. This meant that the 1st Regiment had eight companies totaling 544 rank and file while Woodford's 2nd Regiment had seven companies of 476 rank and file. Colonel Henry, as commander-in-chief of the regular forces, would decide in October which companies were assigned to which regiment.[35]

The regular troops were not the only soldiers ordered to Williamsburg by the Convention. Hundreds of minutemen arrived in the fall to bolster the capital's defense. They comprised a second tier of Virginia's new military establishment. The Convention authorized sixteen battalions of minutemen. These men were drawn from the ranks of the militia and were *"more strictly trained to proper discipline"* than the ordinary militia.[36] Each district was ordered to raise a 500 man battalion of minutemen *"from the age of sixteen to fifty, to be divided into ten companies of fifty men each."*[37]

---

[33] Ibid., 458-459
[34] William W. Hening, ed., *The Statutes at Large Being a Collection of all the Laws of Virginia,* Vol. 9, (Richmond: J. & G. Cochran, 1821), 10, 16
[35] Note: Colonel Henry assigned the companies to their respective regiments on October 21, 1775.
See: Brent Tarter, ed.,"The Orderly Book of the 2nd Virginia Regiment," *Virginia Magazine of History and Biography,* Vol. 85, No. 2 (April 1977), 170-171
[36] Hening, 16
[37] Ibid., 16-17

The last tier of Virginia's new military establishment was the traditional county militia. The Convention decreed that:

> *All male persons, hired servants, and apprentices, above the age of sixteen, and under fifty years...shall be enlisted into the militia...and formed into companies....*[38]

Needless to say, Virginia was awash with military activity in September 1775. County and district committees hurried to appoint officers and issue instructions and the selected officers scrambled to set their personal affairs in order before they reported for duty. Colonel Woodford was no different and one of the things he did before he left his family at Windsor was to write to General George Washington, who was encamped outside of Boston with the recently formed continental army, to solicit his advice.[39] Unfortunately, a copy of Woodford's letter has not been found, but General Washington's reply in November refers to it and includes the advice that Woodford sought.

Washington began by paying Colonel Woodford a compliment. "*I do not mean to flatter, when I assure you, that I highly approve of your appointment.*"[40] Washington then shared a detailed list of suggestions on troop discipline, drill, and preparedness. He concluded the letter warmly:

> *My compliments to Mrs. Woodford; and that every success may attend you, in this glorious struggle, is*

---

[38] Ibid., 27-28
[39] W.W. Abbot, ed. "George Washington to Colonel William Woodford, 10 November, 1775,"" *The Papers of George Washington: Revolutionary War Series,* Vol. 2, (Charlottesville: University Press of Virginia, 1987), 346-347
[40] Ibid.

the sincere and ardent wish of, dear Sir, your affectionate humble servant.[41]

The day after Woodford penned his letter to Washington, the Virginia Committee of Safety, a body created by the 3rd Convention to govern Virginia when the convention was not in session (and headed by Woodford's friend, Edmund Pendleton) met in Hanover Town and authorized two month's pay in advance to Colonel Henry and Colonel Woodford and their fellow staff officers. The committee also issued commissions to the officers who were instructed to report to Williamsburg.[42]

It is uncertain when Colonel Woodford arrived in Williamsburg. He does not appear in the orderly book of the 2nd Virginia Regiment until October 18th, and there is no other reference to him in the weekly gazettes or in personal letters until after that date. Alexander Purdie's *Virginia Gazette* announced the arrival of Colonel Henry in the capital on September 22nd, but Woodford's arrival is a mystery.[43]

If Colonel Woodford did join Colonel Henry in Williamsburg prior to October 18th, he would have found a city abuzz with activity. The task of organizing and regulating the newly arrived troops initially fell to Colonel Thomas Bullit, the adjutant general of Virginia's regular forces. He set

---

[41] Ibid., 347

[42] Robert L. Scribner and Brent Tarter, eds., "Proceedings of the Virginia Committee of Safety, 19 September, 1775," *Revolutionary Virginia: The Road to Independence*, Vol. 4, (University Press of Virginia, 1978), 126 and
Brent Tarter, ed.,"The Orderly Book of the 2nd Virginia Regiment," *Virginia Magazine of History and Biography*, Vol. 85, No. 2 (April 1977), 157

[43] Purdie, *Virginia Gazette*, "22 September, 1775," 2

about in late September to lay out an encampment behind the Wren building at the College of William and Mary and to establish order among the troops. Instructions on rations, arms, ammunition, latrines, guard duty, troop formations, drill, and so on were issued and often repeated in an effort to properly organize the new troops.[44]

Colonel Henry formally took command of the troops assembled in Williamsburg on October 7th.[45] Colonel Woodford assumed his duties in Williamsburg nearly two weeks later.[46] On October 21st, 1775, Colonel Henry assigned the regular companies assembled in the capital to their regiments.[47] With the two regiments of regulars finally formed and organized, the two officers, Henry and Woodford, assumed command of their respective regiments.

Just three days later, Colonel Woodford was ordered by the Committee of Safety (now in Williamsburg) to march the 2nd Regiment and five companies of Culpeper Minutemen to Norfolk to, *"use your best endeavours for protecting and defending the persons and properties of all friends to the cause of America...."*[48]

Before Colonel Woodford could comply with this order, however, a crisis 30 miles south of Williamsburg in the town of Hampton drew him away from his regiment and into Virginia's opening battle of the Revolutionary War.

---

[44] Tarter, ed.,"The Orderly Book of the 2nd Virginia Regiment," *Virginia Magazine of History and Biography*, Vol. 85, No. 2, 159-163
[45] Ibid., 162-168
[46] Ibid., 168-169
[47] Ibid., 170-171
[48] Scribner and Tarter, eds., "Orders for Colonel William Woodford, 24 October, 1775," *Revolutionary Virginia: The Road to Independence*, Vol. 4, 270-271

Map of Hampton

Chapter Three

"Americans will die, or be free!"

Fall 1775

On the morning of October 26th, the inhabitants of Hampton, Virginia awoke to discover a small flotilla of armed tenders (ships) anchored at the mouth of the Hampton River. Captain Matthew Squire, the commander of the 14 gun British sloop of war *H.M.S. Otter*, (anchored off Norfolk) commanded this collection of vessels, which consisted of a large schooner, two smaller sloops, and two even smaller pilot boats, all crewed by a mix of sailors from the *Otter* and loyalist Virginians and runaway slaves.

Captain Squire declared that he intended to burn Hampton in retaliation for an incident that occurred almost two months earlier, the pillaging and destruction of one of the *Otter's* tenders that had washed ashore in a storm (with Captain Squire on it) in early September.

Hampton's defenders, which included, *"a company of regulars and a company of minute-men who had been placed there in consequence of former threats...against* [Hampton], *made the best disposition to prevent their landing, aided by a body of militia, who were suddenly called together on the occasion.*[1] Armed with muskets and more importantly, rifles, (the accuracy of which would play a significant role in the battle to come) the defiant Virginians in Hampton were resolved to stand against the cannon of Captain Squire's ships.

---

[1] Dixon and Hunter, *Virginia Gazette,* "28 October, 1775," 3

Shots rang out as soon as Squire's small flotilla proceeded up the Hampton River towards Hampton. Virginia's Royal Governor John Murray, the Earl of Dunmore, (who was aboard a ship a few miles away in Norfolk) recounted the affair as it was told to him:

> S*ome of the King's tenders went pretty close into Hampton Road. So soon as the rebels perceived them, they marched out against them and the moment they got within shot of our people, Mr. George Nicholas…who commanded a party of rebels at that time at Hampton, fired at one of the tenders, whose example was followed by his whole party. The tenders returned the fire but without the least effect.*[2]

Dunmore laid the blame for the first shot at Hampton (and thus the inauguration of war in Virginia) on Captain George Nicholas of Colonel Woodford's 2nd Virginia Regiment. Nicholas was the son of Virginia's prominent treasurer, Robert Carter Nicholas. An American eyewitness, however, saw the engagement differently. He reported that as the British tenders [ships] approached Hampton,

> *Two vollies of musquetry were discharged from the tenders, and answered by captain Lyne from his post by a rifle, which was answered by a four pounder from one of the tenders; then began a pretty warm fire from all the tenders. Captain Nicholas, observing this, soon joined about 25 of his men. The*

---

[2] K.G. Davis, ed., "Lord Dunmore to Lord Dartmouth, 6 December, 1775 through February, 1776," *Documents of the American Revolution*, Vol. 12, (Irish University Press, 1976), 58

> *fire of our musquetry caused the tender nighest to us to sheer off some distance.*[3]

Captain Lyne of the local minute-man company was identified in a second account as the one, "*who fired the first gun in the attack at the mouth of the river,* [and] *killed a man by that very fire.*"[4]

Although it appears uncertain who fired first at Hampton, the conflict that ensued marked the first engagement of the Revolutionary War in Virginia and lasted for over an hour. Unable to maneuver past several sunken vessels obstructing the harbor, Captain Squire's tenders were raked with rifle and musket fire from shore. Their crews responded with cannon, swivel, and musket fire, but it apparently had little effect on the Virginians. One rebel combatant recalled that,

> *The fire* [from the tenders] *consisted of 4 pounders, grape shot etc. for about an hour. Not a man of our's was hurt. Whether our men did any damage is uncertain. They could not get nigher than 300 yards. Some say they saw men fall in one of the tenders.*[5]

Pinkney's *Virginia Gazette* boasted of the bravery the Virginians displayed against Captain Squire's flotilla:

> *No troops could shew more intrepidity than the raw, new raised men, under the command of captain Nicholas, of the second regiment, and captain Lyne, of the minute men, together with some of the country*

---

[3] Pinkney, *Virginia Gazette*, "2 November, 1775," 2
[4] Pinkney, *Virginia Gazette*, "26 October, 1775," 3
[5] Pinkney, *Virginia Gazette*, "2 November, 1775," 2

> *militia. These brave young officers, at the head of their men, without the least cover or breast-work, on the open shore, stood a discharge of 4 pounders, and other cannon, from a large schooner commanded by captain Squire himself, and from a sloop and two tenders, which played on them with all their guns, swivels, and muskets. They stood cooly till the vessels were near enough for them to do execution, when they began a brisk and well directed fire, which forced the little squadron to retire.*[6]

Yet another account of the battle, apparently published in the evening of the initial engagement, asserted that the defenders of Hampton were eager to face Dunmore's forces again:

> *The troops in town are in high spirits, and wish for* [another] *attack in this quarter; they are all excellent marksmen, and fine, bold fellows. After all the firing at the houses in Hampton, there were only a few windows broke, and a door panel. Lord Dunmore may now see he has not cowards to deal with.*[7]

Hampton's defenders would get their wish for further combat the next morning.

Colonel Woodford Arrives

Although the gunfire at Hampton ended at nightfall, both sides remained active. Under cover of darkness and heavy rain the British tenders quietly moved up to the sunken obstructions and worked to clear a passage through the

---

[6] Pinkney, *Virginia Gazette,* "26 October, 1775," 3
[7] Ibid.

channel while the rebels strengthened their breastworks on the town wharf and anxiously waited for reinforcements from Williamsburg. Colonel Woodford marched all night in the driving rain with a company of Culpeper Minutemen to reach Hampton by morning and assume command of the rebel forces. In a letter to his friend Thomas Jefferson, John Page, a prominent resident of nearby Gloucester County, provided a detailed account (probably supplied by Woodford) of the resumption of combat on October 27th:

> *Col. Woodford accompanied Captain Buford's rifle company through a heavy rain to Hampton and arrived about 7 a.m. When the Col. Entered the Town, having left the Rifle Men in the Church to dry themselves, he rode down to the River, took A view of the Town, and then seeing the Six Tenders at Anchor in the River went to Col. Cary's to dry himself and eat his Breakfast. But before he could do either, the Tenders had cut their Way through the Vessel's Boltsprit which was sunk to impede their Passage and having a very fresh and fair Gale had anchored in the Creek and abreast of the Town.*
>
> *The People were so astonished at their unexpected and sudden Arrival that they stood staring at them and omitted to give the Col. the least Notice of their approach. The first Intelligence he had of this Affair was from the Discharge of a 4 Pounder. He mounted his Horse and riding down to the Warf found that the People of the Town had abandoned their Houses*

and...the Militia had left the Breast Work which had been thrown up across the Wharf and street.[8]

Colonel Woodford deployed Captain Nicholas's company of 2nd Virginians, Captain Abraham Buford's Culpeper riflemen, and the local minute-men and militia amongst the buildings overlooking the harbor and behind breastworks on the shore and at the wharf. John Page's account of the battle continued:

> *The Fire was now general and constant on both Sides. Cannon Balls Grape Shot and Musket Balls whistled over the Heads of our Men, Whilst our Muskets and Rifles poured Showers of Balls into their Vessels and they were so well directed that the Men on Board the Schooner in which Captain Squires himself commanded were unable to stand to their 4 Pounders which were not sheltered by a Netting and gave but one Round of them but kept up an incessant firing of smaller Guns and swivels, as did 2 Sloops and 3 Boats for more than an Hour and ¼ when they slipt their Cables and towed out except the Hawk Tender a Pilot Boat that had been taken some Time before from a Man of Hampton....*
>
> *In her they found 3 wounded Men 6 Sailors and 2 Negroes. Lieut. Wright who commanded her had been forced to jump over Board and was attended to the Shore by 2 Negroes and a white Man, one of the*

---

[8] William Clark, ed., "John Page to Thomas Jefferson, 11 November, 1775," *Naval Documents of the American Revolution,* Vol. 2, (Washington, D.C.: U.S. Printing Office, 1964), 991-992

Negroes was shot by a Rifle Man across the Creek at 400 yds. distance. If Col. Woodford's Men whom he had ordered round to the Creeks Mouth could have got there soon enough they would undoubtedly have taken the little Squadron, for the Sailors could not possibly have towed them through their Fire. Although the nearest of the Tenders was 3 Hundred Yds, and the farthest about 450 from our Men, yet our Fire was so well directed that the Sailors were not able to stand to their Guns and serve them properly but fired them at Random at an Unaccountable Degree of Elevation.[9]

John Page's account of the battle was supported by accounts that appeared in the different Virginia newspapers. One printed in Pinkney's Gazette a week after the engagement reported that,

In the night they cut a passage through the vessels that were sunk, and the next morning, about 8 o' clock (which was about half an hour after colonel Woodford and captain Buford arrived with a rifle company) 5 tenders, to wit, a large schooner, 2 sloops, and 2 pilot boats, passed the passage they had cleared, and drew up a-breast of the town; they then gave 3 cheers, and began a heavy fire.
Colonel Woodford immediately posted captain Nicholas with his company on one side of the main street, and captain Buford with his riflemen on the other, who were joined by the town company of militia;

---

[9] Clark, ed., "John Page to Thomas Jefferson, 11 November, 1775," *Naval Documents of the American Revolution,* Vol. 2, 991-92

captain Lyne with his company [of minute-men] *was ordered to march to the cross roads just out of town to sustain any attack that might come from James or Back river. The colonel had been informed that men were landed from both these rivers. The musquet and rifle balls soon began to fly so thick that few men were seen upon the decks. The engagement continued very warm for some time. At length they began to cut and slip their cables, and all cleared themselves, except one, which was boarded and taken by some of our men. They took in her the gunner and 7 men, 3 of whom were wounded, 2 mortally (both since dead), 1 white woman, and 2 negro men. Lieutenant Wright, who commanded the prize, after receiving a ball, jumped overboard, and it is thought he was not able to reach the tenders. Several more jumped overboard; but it is not known what is become of them, or what damage is done on board of the other tenders. In those 2 different actions, Mr. Printer, officers and soldiers of the regular, minute, and militia acted with a spirit becoming freemen and Americans, and must evince that Americans will die, or be free!*[10]

Battered by heavy and accurate small arms fire from Colonel Woodford's troops in town and along the wharf, Captain Squire reluctantly withdrew out of range, no doubt furious, yet humbled by the resistance he met.

Six months after the bloodshed of Lexington and Concord, warfare had finally erupted in Virginia. In the weeks that

---

[10] Pinkney, *Virginia Gazette*, "2 November, 1775," 2

followed the Battle of Hampton, the conflict intensified significantly.

Woodford Proceeds to Norfolk

Brimming with confidence after his victory at Hampton, Colonel Woodford returned to Williamsburg, eager to lead his regiment and the minutemen attached to him, totaling over 650 men, to Norfolk.[11] A shortage of tents and the sudden arrival of a portion of Lord Dunmore's flotilla off of Burwell's Landing and Jamestown Island (which threatened Williamsburg and obstructed Woodford's passage over the river) delayed Woodford's departure. Skirmishes between Dunmore's ships and rebel parties onshore occurred daily and although they resulted in little loss for either side, they added to Colonel Woodford's delay.

Dunmore's effort to prevent the passage of rebel troops across the James River was not the only factor behind Colonel Woodford's lack of progress southward. His force was also plagued by supply shortages. John Page described the challenges that confronted Colonel Woodford and his men:

> *The Committee had resolved a Month ago to send down the 2d Regimt & the Culpeper Battalion of Minute Men to Norfolk – but for want of Arms, Tents &ct. they were unable to march, the whole of them til the Day before Yesterday* [November 9th] *– A Detachment passed Jas River about 10 days since under Major Spotswood -- & were ordered to halt at*

---

[11] Julian P. Boyd, ed., "Edmund Pendleton to Thomas Jefferson, 16 November, 1775," *The Papers of Thomas Jefferson,* Vol. 1, (Princeton, NJ: Princeton University Press, 1950), 260-61

> *Cobham til joined by the Col. & Remainder of the Forces—This Junction was delayed several Days for want of Necessaries, & several more by high Winds & the Interposition of the King Fisher & several Tenders – which obliged our Men to cross the River higher up than was at first intended – This Delay for Want of proper Arms &ct. has been very mortifying to us, & has proved fatal to our Friends in the Neighborhood of Norfolk.*[12]

By mid-November, Colonel Woodford had gathered sufficient supplies to proceed to Norfolk and he led his troops upriver to cross the James River unmolested.[13]

Although concern remained about the supply situation, Colonel Woodford and his men were confident that they could handle what lay ahead. The Virginia Committee of Safety, in a long report to the Continental Congress, estimated Dunmore's force in mid-November at:

> *The Otter and* [Kingfisher] -- *20 guns & 170 men each; in this number however are included those which man Occasionally the Following tenders Vizt*
>
> *4 Schooners*
> *3 Sloops*
> *3 Pilot Boats*

---

[12] Clark, ed., "John Page to Thomas Jefferson, 11 November, 1775," *Naval Documents of the American Revolution*, Vol. 2, 991-92
[13] Clark, ed., "John Page to Congress, 17 Nov., 1775," *Naval Documents of the American Revolution*, Vol. 2, 1061
  Note: Page provided Congress with a detailed explanation for Woodford's delay in marching south. (See next page)

> *On board these Tenders are some 4 & 3 pounders, besides Swivels.*[14]

The Committee added that three other ships completed Dunmore's flotilla.

In addition to his small flotilla, it was estimated that Dunmore's ground forces amounted to only 300 men, about half of which were British regulars from the 14th Regiment and the other half a mixture of loyalist Virginians and runaway slaves and servants.[15]

The Committee of Safety sheepishly noted, *"the disgraceful patience & Suffering of some of our people,"* in Norfolk and the surrounding area to Dunmore's rule. They also, however, acknowledged that, *the exposed Situation of their Families & property, the want of Arms & Ammunition & their intermixture with Torys, who instead of Assisting were ready every moment to betray them,* accounted for their lack of resistance and resolve.[16] The committee concluded with a confession regarding the vulnerable inhabitants of Norfolk and southern Virginia:

> *We could not protect them, We had men enough, but were left to ransack every corner of the Country for Arms, tents & other necessarys. The few we collected were unavoidably retained here for the protection of our Magazine, Treasury & Records;*[17]

---

[14] Clark, ed., "Virginia Committee of Safety to Congress, 11 November, 1775," *Naval Documents of the American Revolution*, Vol. 2, 993-94
[15] Ibid.
[16] Ibid.
[17] Ibid.

Many believed that the presence of Colonel Woodford and his troops south of the James River would intimidate the supporters of Dunmore and encourage greater resistance or at least non-cooperation to him from the local populace. Thus, it was urgent that Woodford and his men arrive in the region as soon as possible.

Ever since the repulse of Captain Squire's squadron at Hampton, Lord Dunmore longed for a chance to strike back at the rebels. The efforts of his naval squadron off of Jamestown proved ineffectual and only increased his frustration. On November 7th, Dunmore vented some of this frustration in a written proclamation, but he withheld its release out of concern that his own shortage of arms, powder, and troops would not allow him to enforce the proclamation. He needed a significant victory to boost the flagging morale of his few supporters (and attract new supporters) and on November 14th, while Colonel Woodford moved his troops across the James River, Dunmore's troops produced just such a victory, routing a detachment of rebel militia at Kemp's Landing in Princess Anne County (a few miles east of Norfolk).

Dunmore's Force Grows

Lord Dunmore seized upon his victory at Kemp's Landing to issue his proclamation which read in part,

> *I do require every Person capable of bearing Arms, to resort to His Majesty's STANDARD, or be looked upon as Traitors to His Majesty's Crown and Government, and thereby become liable to the Penalty the Law inflicts upon such Offences; such as forfeiture of Life, confiscation of Lands, &c. &c And I do hereby further declare all*

indentured Servants, Negroes, or others (appertaining to Rebels,) free that are able and willing to bear Arms, they joining His Majesty's Troops as soon as may be, for the more speedily reducing this Colony to a proper Sense of their Duty, to His Majesty's Crown and Dignity.[18]

Within days of Dunmore's proclamation hundreds of Virginians answered his call to arms. A resident of Norfolk noted that

The day after [Dunmore hoisted the King's Standard] *the whole Country flocked to it, took the oath of allegiance...and declared their readiness to defend his Majesty's Crown & dignity.... I can assure you that L. Dunmore is so much admired in this part of the County that he might have 500 Volunteers to march with him to any part of Virginia.*[19]

Dunmore informed British General William Howe in Boston of his improved situation:

Immediately on [the victory at Kemp's Landing] *I issued the inclosed Proclamation which has had a Wonderful effect as there are not less than three thousand that have already taken and signed the inclosed Oath.*[20]

---

[18] Clark, ed., "Lord Dunmore's Proclamation," *Naval Documents of the American Revolution,* Vol. 2, 920

[19] Scribner and Tarter, ed., "John Brown, Virginia, to Mr. William Brown, An Intercepted Letter, 21 November, 1775," *Revolutionary Virginia, The Road to Independence,* Vol. 4, 445

[20] Clark, ed., "Lord Dunmore to General William Howe, 30 November, 1775," *Naval Documents of the American Revolution,* Vol. 2, 1209-11

Dunmore was particularly pleased with the impact of the most controversial part of his proclamation, freedom for runaway servants and slaves of rebels who agreed to fight under the King's standard. Dunmore reported to General Howe that

> *The Negroes are flocking in also from all quarters which I hope will oblige the Rebels to disperse to take care of their families, and property, and had I but a few more men here I would March immediately to Williamsburg my former place of residence by which I should soon compel the whole Colony to Submit.*[21]

Dunmore emphasized to Howe his need for arms and supplies, (two items he hoped General Howe might assist him with), and detailed his efforts to organize the large number of runaway slaves and servants and loyal Virginians who had answered his call to arms:

> *We are in great want of small Arms, and if two or three light field pieces and their Carriages could be Spared they would be of great Service to us, also some Cartridge paper of which not a Sheet is to be got here, and all our Cartridges expended.*
>
> *I have...ordered a Regiment (Called the Queens own Loyal Virginia Regiment) of 500 men to be raised immediately consisting...Ten Companys each of which is to consist...50 Privates.*
>
> *You may observe by my Proclamation that I offer freedom to the Slaves, (of all Rebels) that join me, in consequence of which there are between two and three hundred already come in and these I form into a Corps*

---

[21] Ibid.

as fast as they come in giving them white Officers and Non Commissioned Officers in proportion....[22]

Dunmore's success was grudgingly acknowledged by his opponents. John Page informed his friend, Thomas Jefferson,, that Dunmore, "*has made a compleat Conquest of Princess Ann and Norfolk and Numbers of Negroes, and Cowardly Scoundrels flock to his Standard.*"[23]

Page was not completely discouraged by the turn of events, however. He remained hopeful that Colonel Woodford and his force would soon turn the situation around for the rebels:

We hope soon to put a stop to his Career and recover all we have lost, for Col. Woodford after innumerable Delays for want of Arms &c. &c. is by this Time very near him with his Regiment and 250 Minute Men of the Culpeper Batalion and a Number of Volunteers....[24]

---

[22] Ibid.
[23] Boyd, ed., "John Page to Thomas Jefferson, 24 November, 1775," *The Papers of Thomas Jefferson,* Vol. 1, 264-65
[24] Ibid.

Chapter Four

"A Second Bunker's Hill Affair, in Miniature"

Winter 1775

Colonel William Woodford's mixed force of 2nd Virginia regulars and Culpeper minutemen reached Suffolk, 15 miles southwest of Portsmouth, on November 25th.[1] Woodford sent Lieutenant Colonel Charles Scott ahead towards the Great Bridge with over 200 troops to observe the enemy. Ever the cautious officer, Woodford ordered Scott to, *"be safe kept 'till my arrival."*[2]

The Great Bridge was actually a long, narrow, manmade causeway with multiple wooden bridges spanning the southern branch of the Elizabeth River and its tributaries and marshland. Norfolk lay eleven miles north of the main bridge span and since most of the terrain south of Norfolk was marsh and swamp, the Great Bridge road was the primary southern land route to Norfolk.

Lieutenant Colonel Scott was eager to confront Dunmore's small force of *"Tories and Blacks"* (who had removed the planks from the main bridge and were posted in a small

---

[1] D.R. Anderson, ed., "Colonel Woodford to Edmund Pendleton, 26 November, 1775," in "The Letters of Colonel William Woodford, Colonel Robert Howe, and General Charles Lee to Edmund Pendleton," *Richmond College Historical Papers*, (June, 1915), 104
[2] Ibid.

wooden stockade fort on the north bank of the river adjacent to the dismantled bridge) but Colonel Woodford cautioned against such a move.[3] Woodford informed Scott that a severe shortage of ammunition and arms made it impossible for him to march to the Great Bridge until, "*a number of Ball is run, cartridges made, arms Repair'd &ct. &ct.*"[4] The best he could do to support Scott was to send him two more companies of regulars under Major Alexander Spotswood.

In Norfolk, Lord Dunmore and his force of British regulars, runaway slaves, and Tory volunteers, braced for the arrival of Woodford's troops. Dunmore realized that to maintain his base of operation at Norfolk he had to keep control of the Great Bridge. He explained the bridge's importance, and his efforts to defend it, to General Howe:

> *Having heard that a thousand chosen Men belonging to the Rebels, a great part of which were Rifle men, were on their March to attack us here so to cut off our provisions, I determined to take possession of the pass at the great Bridge which Secures us the greatest part of two Counties to supply us with provisions. I accordingly ordered a Stockade Fort to be erected there, which was done in a few days, and I put an Officer and Twenty five men to Garrison it, with some Volunteers and Negroes.*[5]

---

[3] Ibid.
[4] Ibid.
[5] Clark, ed., "Lord Dunmore to General William Howe, 30 November, 1775," *Naval Documents of the American Revolution*, Vol. 2, 1209-11

Bloodshed and Stalemate at the Great Bridge

While Colonel Woodford hurried to make cartridges and equip his troops in Suffolk with functioning firearms, Lieutenant Colonel Scott's advance guard encamped at Great Bridge (just south of the dismantled bridge) and skirmished with Dunmore's forces. The number of casualties that each side claimed was inflicted on the other suggests that the skirmishes were heated.[6]

The bulk of Lieutenant Colonel Scott's force of nearly 200 men was posted behind strong breastworks on the southern edge of the causeway. Sentries were posted forward of these works at night, on what was essentially an island, with the Elizabeth River to their front, a small creek to their rear (fifty yards in front of the main breastworks), and marsh on either side of the causeway. Hidden amongst a few buildings and piles of debris close to the dismantled bridge and Dunmore's fort, Scott's sentries were positioned to alarm their comrades at the breastworks if the enemy approached at night. For their own safety, the sentinels were withdrawn from the island at dawn each day.

Colonel Woodford, with the main body of troops, reached Great Bridge on December 2nd, and informed the Committee of Safety in Williamsburg that,

---

[6] Clark, ed., "Lt. Col. Charles Scott to a Williamsburg Correspondent, 4 December, 1775," *Naval Documents of the American Revolution,* Vol. 2, 1274-75
and
  Davies, ed., "Lord Dunmore to Lord Dartmouth, 6 December through 18 February, 1776," *Documents of the American Revolution,* Vol. 12, 59

> *I...found the Enemy Posted on the opposite side of the Bridge, in a Stockade Fort, with two four pounders, some swivels & wall pieces, with which they keep up a constant Fire, have done no other damage than kill'd Corpl Davis with a cannon ball....*[7]

Woodford estimated that Dunmore's fort was defended by 250 men, most of who were escaped slaves commanded by sergeants of the 14th Regiment.[8] A handful of Tories also helped man the fort. Woodford speculated that it might be possible to capture it, but the presence of cannon meant that its conquest would come at a very high cost in lives:

> *The Enemys Fort, I think, might have been taken, but not without the loss of many of our Men, their Situation is very advantageous, & no way to attack them, but by exposing most of the Troops to their Fire upon a large open Marsh.*[9]

As for his own fortifications, Colonel Woodford reported that

> *We have raised a strong Breastwork upon the lower part of the street joining the Causeway, from which Centries are posted at some old Rubbish not far from the Bridge (which is mostly destroy'd).*[10]

---

[7] Anderson, ed., "Colonel Woodford to Edmund Pendleton, 4 December, 1775," *Richmond College Historical Papers*, 106
[8] Ibid.
[9] Ibid., 107
[10] Ibid.

Although he believed that he held a strong position, Woodford was concerned about his limited supply of gunpowder and the lack of blankets and shoes for his men.

> *Our small stock of Ammunition will be soon expended, & I must request another supply; an additional Blanket to each soldier [will] be very necessary, if to be had. The men are tolerably well at present, but the dampness of this Ground, without straw (which is not to be had) must soon lay many of them up, & Houses that are tolerably safe from the Enemy's Cannon, can only be procured for a few.*[11]

One officer who likely found shelter in one of the "safe" houses out of range (but not earshot) of Dunmore's guns was Lieutenant Colonel Scott. It had been over a week since he had led his detachment ahead of Woodford's main body to the Great Bridge and now that reinforcements had arrived Scott allowed himself to relax. He wrote to a friend on December 4th that

> *Last night was the first of my pulling off my clothes for 12 nights successively. Believe me, my good friend, I never was so fatigued with duty in my whole life;*[12]

Despite the large number of reinforcements, it is likely that Lieutenant Colonel Scott still found it difficult to rest. He explained in his letter that

---

[11] Ibid., 108-09

[12] Clark, ed., "Lt. Col. Charles Scott to a Williamsburg Correspondent, 4 December, 1775," *Naval Documents of the American Revolution*, Vol. 2, 1274-75

> *We still keep up a pretty heavy fire between us, from light to light. We have only lost two men, and about half an hour ago one of our people was shot through the arm, which broke the bone near his hand.*[13]

The skirmishing continued downriver as well. Within days of his arrival, Colonel Woodford sent a large detachment of troops under Colonel Edward Stevens of the Culpeper Minutemen a few miles downriver to strike an enemy guard post on the river. Woodford reported that

> *They crossed about midnight, & got to the Enemy's centinals without being discover'd, one of them challenged & not being answer'd, Fired at our party, the fire was returned by our men, & an over Eagerness at first, & rather a backwardness afterwards, occasion'd some confusion, & prevented the Colo's plan from being so well executed as he intended, however, he [burned] their Fortification & House, in which one negro perished, killed one dead upon the spott, & took two others prisoners...this party (consisting of 26 Blacks & 9 Whites) escaped under the cover of night, he also took four new Muskets.*[14]

Although the bulk of the enemy guard escaped, their post was destroyed. Discovered among the captured troops and abandoned equipment were altered musket balls designed to split into quarters upon impact. Colonel Woodford was

---

[13] Ibid.
[14] Anderson, ed., "Colonel Woodford to Edmund Pendleton, 5 December, 1775," *Richmond College Historical Papers*, 110

outraged by the discovery and sent one to Williamsburg for the Committee of Safety and 4th Virginia Convention to see:

> *The bearer brings you one of the Balls taken out of the cartridges found upon the negro Prisoners, as they were extremely well made, & no doubt by some of the non comd. Officers of the Regulars, will submit it to the Convention, by who's orders this Horrid preparation was made for the Flesh of our Countrymen, the others are prepared in the same manner, likewise all that have been found in the Houses &ct; – I have never suffer'd a soldier of mine to do a thing of this kind – nor will I allow it to be done for the future, notwithstanding this provocation....*[15]

Two nights later, Woodford's troops struck again, attacking the same post – re-occupied and reinforced by Dunmore with 70 men. This time Lieutenant Colonel Scott led the attack with 150 men. Colonel Woodford described the engagement to Edmund Pendleton:

> *I have the pleasure to inform you...that my detachment last night under the Command of Lieut. Colo. Scott beat up the Quarters of the Enemys other party, who I inform'd you had again taken post opposite our Boat Guard, they killed one white man & three negro's, took three of the Latter Prisoners, two of which are wounded, one mortally, with six Muskets & 3 Bayonetts.*[16]

---

[15] Anderson, ed., "Colonel Woodford to Edmund Pendleton, 5 December, 1775," *Richmond College Historical Papers*, 112
[16] Anderson, ed., "Colonel Woodford to Edmund Pendleton, 7 December, 1775," *Richmond College Historical Papers*, 114

Woodford explained that bad luck prevented his troops from surprising the enemy:

> [Colonel Scott] *unluckily fell in with a cart coming from Norfolk, guarded by four men, some distance from the Enemy's post, who Fired upon our party & alarm'd them, otherways there is no doubt most of their men would have fallen into our Hands, their number 70, Scott's party, 150, who all escaped unhurt, one man only was grazed by a Ball in the Thumb.*[17]

Williamsburg Grows Anxious

Although Colonel Woodford's troops had successfully engaged Dunmore's forces in a number of skirmishes, apprehension grew among the leaders in Williamsburg that time was running out to drive Dunmore from Norfolk. Thomas Ludwell Lee of Stafford County summarized the concern of many in the 4th Virginia Convention and Committee of Safety:

> *Our Army has been for some time arrested in its march to Norfolk by a redoubt, or stockade, or hog pen, as they call it here, by way of derision, at the end of this bridge. Tho,' by the way, this hog pen seems filled with a parcel of wild-boars, which we appear not overfond to meddle with. My apprehension is that we shall be amused at this outpost, until Dunmore gets the lines at Norfolk finished; where he is now entrenching,*

---

[17] Ibid.

> & mounting cannon, some hundreds of negro's being employ'd in the work.[18]

Fortunately for Colonel Woodford, reinforcements from North Carolina with cannon were reportedly on the way. Woodford forwarded this news to the Committee of Safety on December 4th

> *They inform me I might expect 4 or 500 men with some Cannon & ammunition at this place tonight, & that they had 900 men at different places in Motion to join us.*[19]

Woodford also reported that his troops

> *Were now making the necessary preparations to raise Batterys for these Cannon upon the most advantageous Ground to play upon their Fort & send a large detachment at the same time to intercept their Retreat.*[20]

Even Lord Dunmore believed that a large reinforcement of rebel troops were on their way to the Great Bridge and this belief helped spur him to action:

> *The Rebels had procured some Cannon from North Carolina, [which were expected to arrive any day] and that they were also to be reinforced from Williamsburg, and knowing that our little Fort was not in a Condition to withstand anything heavier than*

---

[18] William Clark, ed., "Thomas Ludwell Lee to Richard Henry Lee, 9 December, 1775," *Naval Documents of the American Revolution*, Vol. 3, (Washington, D.C.: U.S Printing Office, 1968), 26-27
[19] Anderson, ed., "Colonel Woodford to Edmund Pendleton, 4 December, 1775," *Richmond College Historical Papers*, 108
[20] Ibid.

> *Musquet Shot, I thought it advisable to risqué Something to save the Fort.*[21]

Dunmore Decides to Attack

Concern over rebel cannon was not the only reason Lord Dunmore decided to forsake his fortified position and attack Colonel Woodford's works on December 9th. Dunmore was also influenced by faulty intelligence that he received from the rebel camp. Colonel Woodford explained the importance of this occurrence to the Committee of Safety following the battle:

> *A servant belonging to major* [Thomas] *Marshal, who deserted the other night from col.* [Charles] *Scott's party, has completely taken his lordship in. Lieutenant* [John] *Batut...informs, that this fellow told them not more than 300 shirtmen were here; and that imprudent man* [Dunmore] *catched at the bait, dispatching capt. Leslie with all the regulars (about 200) who arrived at the bridge about 3 o' clock in the morning.*[22]

---

[21] Clark, ed., "Lord Dunmore to Lord Dartmouth, 13 December, 1775," *Naval Documents of the American Revolution,* Vol. 3, 140-41

[22] Clark, ed., "Colonel Woodford to Edmund Pendleton, 9 December, 1775," *Naval Documents of the American Revolution*, Vol. 3, 28

Note: Colonel Woodford repeated this account in a second letter to Edmund Pendleton the next day. A similar account was included in the *Annual Register for the Year 1776*, p. 29

"*It has been said, that we were led into this unfortunate affair, through the designed false intelligence of a pretended deserter, who was tutored for the purpose.*"

It is uncertain whether the servant purposefully or accidently misled Dunmore about Woodford's troop strength, but it appears that Dunmore viewed the news that so few rebels faced him across the causeway at that moment as an opportunity that would soon disappear when the expected reinforcements arrived from North Carolina. As a result, on the evening of December 8th, Dunmore rushed his own reinforcements, including most of the regulars of the 14th Regiment (approximately 120 under Captain Samuel Leslie) as well as a detachment of sailors (to help man the fort's cannon) and about 60 Tory volunteers, from Norfolk to the fort at the Great Bridge.[23] A British midshipman from the *HMS Otter* who participated in the battle recalled that

> *Our troops, with about sixty Townsmen from Norfolk, and a detachment of Sailors from the ships, among whom I had the honour to march, set out from Norfolk to attack once more the Rebels at the great bridge....We arrived at the Fort half an hour after three in the morning, and, after refreshing ourselves, prepared to attack the Rebels in their intrenchments.*[24]

---

[23] Clark, ed., "Letter from a Midshipman on Board H.M. Sloop Otter, 9 December, 1775," *Naval Documents of the American Revolution*, Vol. 3, 29
[24] Ibid.

Battle of Great Bridge

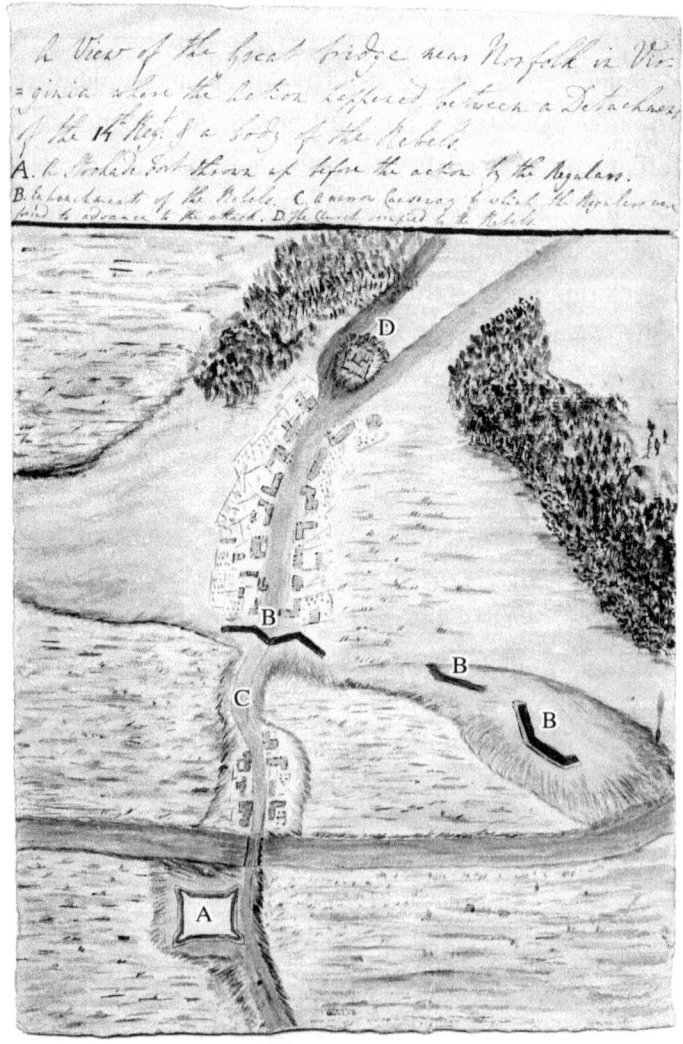

A view of the Great Bridge near Norfolk in Virginia where the action happened between a detachment of the 14th Reg^t & a body of the rebels.
A. A stockade port thrown up before the action by the Regulars.
B. Entrenchment of the Rebels. C. A narrow causeway by which the Regulars were forced to advance to the attack. D. The church occupied by the Rebels.

Courtesy of William L. Clements Library, University of Michigan

These reinforcements joined the garrison of Tories, runaway slaves, and handful of regulars already at the fort early in the morning of December 9th.

Dunmore's Plan

Dunmore hoped to disperse the rebels at Great Bridge before they could be reinforced and instructed Captain Leslie to attack them early in the morning of December 9th. In his report to Lord Dartmouth after the battle, Dunmore explained that his plan called for,

> *Two Companies of Negroes to make a detour,* [cross the river] *and fall in behind the Rebels a little before break of Day in the morning, and just as Day began to break, to fall upon the rear of the Rebels, which* [Dummore] *expected would draw their attention, and make them leave the breast work they had made near the Fort,* [Captain Leslie] *was then with the Regulars, the Volunteers and some recruits to sally out of the Fort, and attack* [the rebel] *breast work*....[25]

Dunmore hoped that the distraction caused by his black troops would allow his main force under Captain Leslie to cross the Elizabeth River, advance along the narrow causeway and storm the rebel breastworks against limited opposition. Unfortunately for Dunmore, miscommunication, or perhaps a misunderstanding of orders, prevented the diversionary attack from occurring. Dunmore noted after the battle that

---

[25] Clark, ed., "Lord Dunmore to Lord Dartmouth, 6 December through 18 February, 1776," *Naval Documents of the American Revolution*, Vol. 3, 141

> *The Negroes by some mistake were sent out of the Fort to guard a pass, where it was thought the Rebels might attempt to pass, and where in fact some of them had Crossed a Night or two before, burnt a house or two, and returned; Captain Leslie not finding the Negroes there, imprudently Sallied out of the Fort at break of Day in the morning....*[26]

The Attack

Under cover of the dim light of dawn Captain Leslie's force of approximately 350 men advanced from their fort and hastily re-laid the bridge planks that had been removed weeks earlier.[27] If the handful of sleepy rebel pickets sheltered by the buildings on the island initially failed to notice the activity at the bridge, the discharge of the fort's cannon undoubtedly drew their attention that way. Startled at what they saw, the rebel sentries opened fire upon Dunmore's troops. One rebel account of the battle included high praise for the sentries:

> *The conduct of our sentinels I cannot pass over in silence. Before they quitted their stations they fired at least three rounds as the enemy were crossing the bridge, and one of them, who was posted behind some shingles, kept his ground till he had fired eight*

---

[26] Ibid.
　　Note: Lord Dunmore claimed in this letter that he left the discretion of whether to actually launch the attack with Captain Leslie.

[27] Clark, ed., "Letter to John Pinkney, 20 December, 1775," *Naval Documents of the American Revolution*, Vol. 3, 186-89

times, and after receiving a whole platoon, made his escape over the causeway into our breast works.[28]

As the handful of sentinels scurried back to the rebel earthworks 150 yards to the rear, their comrades behind the breastwork began to stir, realizing that the gunfire they heard was not the normal morning salute of the past few days.

Four hundred yards south of the earthworks at the main rebel encampment, however, few of Colonel Woodford's troops, who had just been awakened by reveille, took notice of the distant gunfire. Major Alexander Spotswood recalled

We were alarmed this morning by the firing of some guns after reveille beating, which, as the enemy had paid us this compliment several times before, we at first concluded to be nothing but a morning salute.[29]

Colonel Woodford had a similar reaction, recalling that,

After reveille beating, two or three great guns, and some musquetry were discharged from the enemy's fort, which, as it was not an unusual thing, was but little regarded.[30]

The situation was much different at the rebel breastworks. Realizing that they were under attack, the commander of the guard, Lieutenant Edward Travis ordered his small

---

[28] Ibid.
   Note: The brave sentinel who stood his ground for so long was twenty year old Billy Flora, a free born black volunteer from Norfolk.
[29] Peter Force, ed., "Major Spotswood to a Friend in Williamsburgh, 9 December, 1775," *American Archives, Fourth Series,* Vol. 4, (Washington, D.C., U.S. Congress, 1858-1853), 224
[30] Clark, ed., "Col. Woodford to Edmund Pendleton, 10 December, 1775," *Naval Documents of the American Revolution,* Vol. 3, 39-4

detachment of approximately sixty men, *"to reserve their fire till the enemy came within the distance of fifty yards."*[31] A small stream lay about 50 yards in front of the rebel breastworks and served as an excellent range marker for the rebels. To their front across the narrow 150 yard causeway were more than five times their number of enemy troops with two cannon that one rebel recalled were, *"planted on the edge of the island, facing the left of our breast-work,* [and] *played briskly...upon us."*[32]

Joining the cannon at the edge of the island were the Tory and Black soldiers of Dunmore, over 200 strong. Behind them rose the smoke of several buildings -- formerly the outposts of the rebel sentries but now torched by Dunmore's troops. Captain Leslie remained on the island with the Tory and Black troops while Captain Charles Fordyce led the British regulars of the 14th Regiment, 120 strong in a column six abreast, across the narrow causeway to storm the rebel earthworks.[33]

Back in the main "rebel" camp, the gravity of the situation had finally become apparent. Major Spotswood recalled

> *I heard Adjutant Blackburn call out, Boys! stand to your arms! Colonel Woodford and myself immediately got equipped, and ran out; the Colonel pressed down to the breastwork in our front, and my alarm-post being two hundred and fifty yards in another quarter, I*

---

[31] Force, ed., "Major Spotswood to a Friend in Williamsburgh, 9 December, 1775," *American Archives*, Vol. 4, 224 and Clark, "Letter to Pinkney, 20 December, 1775", *Naval Documents of the American Revolution*, Vol. 3, 186-89

[32] Clark, "Letter to Pinkney, 20 December, 1775," *Naval Documents of the American Revolution*, Vol. 3, 186-89

[33] Clark, "Letter from a Midshipman on Board H.M. Sloop Otter, 9 December, 1775" *Naval Documents of the American Revolution*, Vol. 3, 29

> *ran to it as fast as I could, and by the time I had made all ready for engaging, a very heavy fire ensued at the breastwork, in which were not more than sixty men;*[34]

The heavy fire that Major Spotswood heard came from Lieutenant Travis's guard detail and a few brave reinforcements who had rushed forward at the first alarm. Lieutenant John Marshall of the Culpeper Minutemen (and future Chief Justice of the Supreme Court) was at Great Bridge and remembered

> *As is the practice with raw troops, the bravest rushed to the works, where, regardless of order, they kept up a heavy fire on the front of the British column.*[35]

The valor of some of the rebels was also acknowledged by Major Spotswood, who proudly noted in a letter immediately after the engagement that as the redcoats approached the breastworks with fixed bayonets, *"Our young troops received them with firmness, and behaved as well as it was possible for soldiers to do."*[36] In his own letter after the battle, Colonel Woodford also commented on the rebel fire from the breastwork, writing that, *"perhaps a hotter fire never happened, or a greater carnage, for the number of troops."*[37]

---

[34] Force, ed., "Major Spotswood to a Friend in Williamsburgh, 9 December, 1775," *American Archives*, Vol. 4, 224

[35] John Marshall, *The Life of George Washington,* Vol. 2, (Fredericksburg, VA: The Citizens Guild of Washington's Boyhood Home, 1926), 132

[36] Force, ed., "Major Spotswood to a Friend in Williamsburgh, 9 December, 1775," *American Archives*, Vol. 4, 224

[37] Clark, ed., "Col. Woodford to Edmund Pendleton, 10 December, 1775," *Naval Documents of the American Revolution,* Vol. 3, 39-49

The hot fire delivered upon the British originated not only from the breastworks directly in front of them, but also from breastworks to the west of the causeway. Riflemen from the Culpeper Minute Battalion manned this position and poured deadly enfilade fire into the British column's right flank.[38] According to one American account, the intense rebel small arms fire from both positions

> *Threw* [the advancing British regulars] *into some confusion, but they were instantly rallied by a Captain Fordyce, and advanced along the causeway with great resolution, keeping up a constant and heavy fire as they approached. The brave Fordyce exerted himself to keep up their spirits, reminded them of their ancient glory, and waving his hat over his head, encouragingly told them the day was their own. Thus pressing forward, he fell within fifteen steps to the breast-work. His wounds were many, and his death would have been that of a hero, had he met it in a better cause.*[39]

A British participant in the battle noted that

> [The rebel] *fire was so heavy, that, had we not retreated as we did, we should every one have been cut off. Figure to yourself a strong breast-work built across a causeway, on which six men only could advance a-breast; a large swamp almost surrounding them, at the*

---

[38] *The Annual Register for the Year 1776*, 4th ed. 29
[39] Clark, "Letter to Pinkney, 20 December, 1775", *Naval Documents of the American Revolution*, Vol. 3, 186-89

> back of which were two small breast-works to flank us in
> our attack on their intrenchments. Under these disadvantages it was impossible to succeed; yet our men were so enraged, that all the intreaties, and scarcely the threats of their Officers, could prevail on them to retreat; which at last they did.[40]

Captain Fordyce, riddled with buck and ball, was one of many redcoats to fall before the rebel earthworks. Strewn about the ground just a few paces from the Virginians were over thirty British dead and wounded. One rebel officer described a scene of bloody carnage before the breastworks:

> The scene, when the dead and wounded were bro't off, was too much; I then saw the horrors of war in perfection, worse than can be imagin'd; 10 and 12 bullets thro' many; limbs broke in 2 or 3 places; brains turning out. Good God, what a sight![41]

Captain Fordyce and twelve British privates lay dead in front of the breastworks and nearly a score of wounded redcoats, including Lieutenant John Batut, who led the British advance guard, were taken prisoner. An American observer noted that

> The progress of the enemy was now at an end; [the survivors] retreated over the causeway with precipitation, and were dreadfully galled in their rear. Hitherto, on our side only the guard, consisting

---

[40] Clark, "Letter from a Midshipman on Board H.M. Sloop Otter, 9 December, 1775," *Naval Documents of the American Revolution*, Vol. 3, 29

[41] Charles Campbell, ed., "Richard Kidder Meade to Theodorick Bland Jr., 18 December, 1775" *The Bland Papers*, Vol. 1, (1840) 38-39

> *of twenty five, and some others, upon the whole, amounting to not more than ninety, had been engaged. Only the regulars of the 14th regiment, in number one hundred and twenty, had advanced upon the causeway, and about two hundred and thirty tories and negroes had, after crossing the bridge, continued upon the island.*[42]

Although the British assault had been repulsed, the battle was yet over, for Captain Leslie rallied his men on the island:

> *The regulars, after retreating along the causeway, were again rallied by captain Leslie, and the two field pieces continued to play upon our men.*[43]

While Dunmore's troops re-grouped around their cannon, Colonel Woodford led troops from the main camp through heavy artillery fire to reinforce the breastworks:

> *It was at this time that colonel Woodford was advancing down the street to the breast-work with the main body, and against him was now directed the whole fire of the enemy. Never were cannon better served, but yet in the face of them and the musquetry, which kept up a continual blaze, our men marched on with the utmost intrepidity.*[44]

Major Spotswood also noted the severity of the enemy cannon fire:

---

[42] Clark, "Letter to Pinkney, 20 December, 1775," *Naval Documents of the American Revolution*, Vol. 3, 186-89
[43] Ibid.
[44] Ibid.

> *The [enemy] field pieces raked the whole length of the street, and absolutely threw double-headed shot as far as the church, and afterwards, as our troops approached, cannonaded them heavily with grapeshot.*[45]

Spotswood credited divine providence for protecting all but one man, who was wounded in the hand, from the intense artillery barrage.[46]

With Dunmore's battered troops stubbornly remaining on the island, Colonel Woodford sent Colonel Edward Stevens with the Culpeper minutemen to reinforce the riflemen on the left flank. The rebel militia poured more deadly enfilade fire from their rifles upon Captain Leslie's troops. The accurate American rifle fire finally prompted Captain Leslie, who was dismayed at his losses (especially that of his nephew, Lieutenant Peter Leslie) to withdraw to the fort. One "rebel" noted that

> *The enemy fled into their fort, leaving behind them the two field pieces, which, however, they took care to spike up with nails. Many were killed and wounded in the flight, but colonel Woodford very prudently restrained his troops from urging their pursuit too far. From the beginning of the attack till the repulse from the breast work might be about fourteen or fifteen minutes; till the total defeat upwards of half an hour. It is said that some of the enemy preferred death to captivity, from fear of being scalped, which lord Dunmore inhumanly told*

---

[45] Force, ed., "Major Spotswood to a Friend in Williamsburgh, 9 December, 1775," *American Archives*, Vol. 4, 224
[46] Ibid.

them would be their fate should they be taken alive. Thirty one, killed and wounded, fell into our hands, and the number borne off was much greater.*[47]

Aftermath

The Battle of Great Bridge was a decisive victory for the Virginians. Colonel Woodford proudly described it as, *"a second Bunker's Hill affair, in miniature; with this difference, that we kept our post, and had only one man wounded in the hand."*[48] The British 14th Regiment of Foot, on the other hand, was shattered in the attack. Their brave, bold assault on the rebel breastworks cost them half their men. Colonel Woodford initially estimated Dunmore's losses at 50 men, noting that some of their dead and wounded were taken back to the fort. He reported that

> *We buried 12, besides...*[Captain Fordyce] *(him with all the military honors due to his rank) and have prisoners lieutenant Batut, and 16 privates; all wounded; 35 stands of arms and accoutrements, 3 officers* [fusils], *powder, ball, and cartridges, with sundry other things, have likewise fallen into our hands.*[49]

Dunmore's report on the 14th Regiment's losses (which was presumably more accurate) claimed 3 officers and 17 men killed and 1 officer and 43 men wounded.[50] The number of

---

[47] Clark, "Letter to Pinkney, 20 December, 1775," *Naval Documents of the American Revolution*, Vol. 3, 186-89
[48] Clark, ed., "Colonel Woodford to Edmund Pendleton, 10 December, 1775," *Naval Documents of the American Revolution*, Vol. 3, 39-40
[49] Ibid.
[50] Clark, ed., "Lord Dunmore to Lord Dartmouth, 13 December, 1775" *Naval Documents of the American Revolution*, Vol. 3, 141

casualties among Dunmore's Tory and black soldiers is unknown.

Calm settled over the causeway soon after the battle as Colonel Woodford dispatched an officer under a flag of truce to allow Captain Leslie to collect his dead and wounded from the battlefield.[51] One observer reported that

> *The work of death being over, every one's attention was directed to the succor* [assistance] *of the unhappy sufferers, and it is an undoubted fact, that captain Leslie was so affected with the tenderness of our troops towards those who were yet capable of assistance, that he gave signs from the fort of his thankfulness for it.*[52]

With both sides secure behind their fortifications the sun set with no more fighting. Captain Leslie abandoned the fort shortly after dark and marched the entire garrison to Norfolk. Lord Dunmore explained Captain Leslie's decision in a letter a few days later:

> *This loss having so much weakened our before but very weak Corps, and Captain Leslie being much depressed by the loss of Lieutenant Leslie, his Nephew, and thinking that the Enemy elated with this little advantage they had gained over us, might force their way across the branch, either above, or below, and by that means, Cut off the Communication between us,*

---

[51] Clark, ed., "Colonel Woodford to Edmund Pendleton, 9 December, 1775," *Naval Documents of the American Revolution*, Vol. 3, 28

[52] Clark, "Letter to Pinkney, 20 December, 1775," *Naval Documents of the American Revolution*, Vol. 3, 186-89

> *determined to evacuate the Fort, and accordingly left it soon after it was dark, and returned with the whole to this place* [Norfolk]; *The Rebels however remained at the Bridge for a day or two.*[53]

Colonel Woodford's troops took possession of the abandoned fort in the morning and found it in disarray. Woodford reported that

> *We...found therein* the fort] *the stores mentioned in the enclosed list, to wit, 7 guns, 4 of them sorry, 1 bayonet, 29 spades, 2 shovels, 6 cannon, a few shot, some bedding, a part of a hogshead of rum, two or more barrels, the contents unknown, but supposed to be rum, 2 barrels of bread, about 20 quarters of beef, half a box of candles, 4 or 5 dozen of quart bottles , 4 or 5 iron pots, a few axes and old lumber; the spikes, I find, cannot be got out of the cannon without drilling.*[54]

Woodford made another observation that led him to believe the enemy had suffered much greater than he realized:

> *From the vast effusion of blood on the bridge, and in the fort, from the accounts of the sentries, who saw many bodies carried out of the fort to be interred and other circumstances, I conceive their loss to be much greater than I thought it yesterday, and the victory to be complete.*[55]

---

[53] Clark, ed., "Lord Dunmore to Lord Dartmouth, 13 December, 1775," *Naval Documents of the American Revolution,* Vol. 3, 140-41

[54] Clark, ed., "Colonel Woodford to Edmund Pendleton, 10 December, 1775," *Naval Documents of the American Revolution,* Vol. 3, 40-41

[55] Ibid.

The 14th Regiment's heavy losses apparently had a strong impact on their commander, Captain Leslie. The Virginia Committee of Safety in Williamsburg gleefully reported a few days after the battle that

> *The Regulars, disgusted, refused to fight in junction with Blacks; and Captain Leslie, we are told, declared no more of his troops should be sacrificed to whims, and put them on board the ships, in consequence of which Norfolk is abandoned, and we expect is now occupied by our troops, who were on their march there when our last account was dispatched.*[56]

---

[56] Clark, ed., "Letter from the Virginia Committee of Safety, 16 December, 1775," *Naval Documents of the American Revolution*, Vol. 3, 132

Chapter Five

"I Would Wish to Keep up the Greatest Harmony Between Us"

1776

Captain Leslie and Lord Dunmore were not the only ones at odds with each other, a dispute over rank and the chain of command between Colonel Woodford, Colonel Henry, and the Committee of Safety and Convention had simmered for weeks and manifested itself into insubordinate behavior among the troops. Colonel Henry first addressed the issue through general orders on October 28th, nearly six weeks before the Battle of Great Bridge.

> *The Soldiers of the first Regiment are to shew the Officers of the 2d Regiment the same Regard & Respect, As the Officers of their own Regiment, of Like Rank. The Soldiers of the 2nd Regiment are to observe the same Conduct of the 1st Regiment. This is done to abolish all party Distinctions.*[1]

---

[1] Brent Tarter, ed., "The Orderly Book of the Second Virginia Regiment, September 27, 1775 – April 15, 1776, General Orders, 28 October, 1775," *Virginia Magazine of History and* Biography, Vol. 85, No. 2, (April 1977), 175

The Committee of Safety's decision four days earlier to send Colonel Woodford and the 2nd Regiment, along with the Culpeper Minute Battalion, to Norfolk instead of Colonel Henry and the 1st Regiment, likely contributed to tension between the two regiments and was not greeted well by Henry and his officers in the 1st Regiment. These officers met in early November without Colonel Henry or any officer of the 2nd Regiment present to discuss their grievances. They claimed that they did so out of concern that Woodford's departure to Norfolk would leave Williamsburg inadequately supplied and defended, but the Committee of Safety challenged this contention and raised their own concern over the conduct of these officers with Colonel Henry:

> *The Committee of Safety think the Council of officers without your presence irregular, & look upon it as a mark of their Suspicion of our Judgment and prudence in providing for the Safety of this place.... It also appears extraordinary that a Council of Officers, if it may be called so, on a matter of such importance to the Country* [of the defense of the capital] *should be held without admitting the Commander in Chief, the Colonel or a single Officer of the 2d Regiment.*[2]

The dispute escalated considerably in early December when Colonel Henry grew annoyed at Colonel Woodford's lack of communication with him about the situation Woodford faced at Great Bridge. Henry believed that as the commander in chief of Virginia's forces, Colonel Woodford was obliged to report to Henry on a regular basis via dispatches from Great

---

[2] Scribner and Tarter, eds., "Vice-President John Page to Patrick Henry, 4 November, 1775," *Revolutionary Virginia*, Vol. 4, 321

Bridge. Woodford initially seemed to agree, explaining in a dispatch to the Committee of Safety on November 26th that he understood that Colonel Henry had returned to his home so, *"I have therefore not wrote him, which I should have consider'd it my Duty to have done, had he been at WmsBurg."*[3] Woodford asked the committee to share the contents of the letter with Henry upon his return and to, *"make an Apology for my not writing."*[4]

Henry was apparently not pleased at what he viewed as Woodford's violation of the chain of command. As commander in chief, Henry believed that Colonel Woodford was obliged to report directly to him, not indirectly through the Committee of Safety.[5] When nearly two more weeks passed without any direct word from Colonel Woodford, Henry wrote to Woodford, but in doing so, the normally direct Henry only hinted at his annoyance:

> *Not hearing of any dispatch from you for a long time, I can no longer forbear sending to know your situation, and what has occurred. Every one, as well as myself, is vastly anxious to hear how all stands with you. In case you think anything could be done to aid and forward the enterprise you have in hand, please to write it. But I wish to know your situation*

---

[3] Scribner and Tarter, eds., "Colonel William Woodford to John Page, Vice President of the Virginia Committee of Safety, 26 November, 1775," *Revolutionary Virginia*, Vol. 4, 478-479
[4] Ibid.
[5] William Wirt Henry, *Patrick Henry: Life, Correspondence and Speeches*, Vol. 1, (New York: Charles Scribner's Sons, 1891), 341

particularly, with that of the enemy, that the whole may be laid before the convention now here.[6]

Colonel Woodford received Henry's letter the next day via a series of express riders. The Convention, which convened on December 4th, wished to keep in daily contact with Woodford and authorized him to employ a secretary, similar to one granted Colonel Henry as commander in chief. This undoubtedly annoyed Colonel Henry further as it suggested that Woodford was of an equal rank to Henry.

Colonel Woodford's reply to Henry conveyed a similar view and brought the whole dispute to a head. Woodford first claimed that he had not bothered to write to Henry because it was Woodford's understanding that Henry was not in Williamsburg. Woodford added that he assumed that the convention would share the contents of his dispatches with Henry, "*as commanding officer of the troops in Williamsburg.*"[7] Woodford's choice of words was interesting as he clearly suggested that Henry did not have command of troops outside of Williamsburg (such as the ones at Great Bridge). Just in case this was not clear, Woodford declared in his next sentence that

> *When joined, I shall always esteem myself immediately under your command, and will obey accordingly; but when sent to command a separate and distinct body of troops, under the immediate*

---

[6] Robert Scribner and Brent Tarter, eds., "Colonel Patrick Henry On Virginia service to William Woodford, esq., colonel of the second regiment of the Virginia forces, 6 December, 1775," *Revolutionary Virginia*, Vol. 5, (University Press of Virginia, 1979), 68

[7] Scribner and Tarter, eds., "Colonel William Woodford to Colonel Patrick Henry, 7 December, 1775," *Revolutionary Virginia*, Vol. 5, 77

instructions of the committee of safety –whenever that body or the honourable convention is sitting, I look upon it as my indispensable duty to address my intelligence to them, as the supreme power in this colony. If I judge wrong, I hope that honourable body will set me right.[8]

Woodford must have known that his words were a slap in Henry's face and his claim that, "*I would wish to keep up the greatest harmony between us, for the good of the cause we are engaged,*" was likely viewed by Henry as insincere.[9]

Colonel Woodford was, however, on solid ground with his interpretation of the command arrangement of the Virginia troops. Colonel Henry's commission did indeed grant him the title of colonel of the first regiment and commander in chief, but it was "*commander in chief of all such other forces as may, by order of the Convention, or Committee of Safety, be directed to act in conjunction with them.*"[10] Colonel Woodford could argue that his assignment at Great Bridge was a separate command, not in conjunction with Colonel Henry, and therefore, not under his direct command.

The Committee of Safety agreed with Woodford, offering Colonel Henry just a small face saving token when it

Resolved Unanimously that Colonel Woodford, altho acting upon separate or detached command, ought to correspond with Colonel Henry & make returns to him at proper times of the state & condition of the forces

---

[8] Ibid.
[9] Ibid.
[10] Scribner and Tarter, eds., "Footnote 5 " *Revolutionary Virginia*, Vol. 4, 125

> *under his command & to be subject to his orders, when the Convention or Committee of safety are not sitting, but that whilst either of these bodies are sitting he ought to receive his Orders from one of them.*[11]

Since it was very unlikely that there would be an instance that neither the Convention or Committee of Safety was not in session, the resolve essentially removed Colonel Henry's authority over Colonel Woodford and the 2nd Virginia Regiment, making Henry a figurehead commander in chief.

Colonel Henry was not alone in his loss of authority, however. The arrival of Colonel Robert Howe of North Carolina (who held a commission from the Continental Congress) just a few days after the battle of Great Bridge meant that Colonel Woodford had to defer to Howe. Woodford acknowledged Howe's authority in a letter to the Committee of Safety and assured the committee that he would work in harmony with Colonel Howe:

> *There is so good an understanding between that Gentn [Howe] & myself, that the Convention need be under no apprehension of a disagreement.*[12]

For his part, Colonel Howe, in a dispatch to the Virginia Convention, was extremely complimentary of Woodford and expressed some reluctance at assuming command.

---

[11] Scribner and Tarter, eds., "An Instruction to Colonel Woodford, 22 December, 1775," *Revolutionary Virginia*, Vol. 5, 221

[12] D.R. Anderson, ed., "The Letters of Col. Wm. Woodford…to Edmund Pendleton, 14 December, 1775," *Richmond College Historical Papers*, June, 1905, 127

It is with diffidence, sir, that I undertake this charge; and I must add however honourable, with reluctance, as I supersede a gentleman I so much esteem, whose abilities I know to be equal to the duties of the station, and who hath so amply filled the measure of his duty.[13]

The two officers appeared to work well together and turned their attention to Norfolk, sending a joint dispatch to the city's leaders and inhabitants on December 14th, promising no harm to those who cooperated with them.

Dunmore's Floating Town

Lord Dunmore's decisive defeat at Great Bridge sparked panic among his supporters at Norfolk and within a week of the battle most had fled with as many valuables as they could take to an array of ships anchored in the harbor. One Tory, who sought refuge aboard the H.M.S. *Kingfisher*, noted that

> *This unfortunate attack* [at the Great Bridge] *which was made in the morning about sunrise dispirited most people.... All thoughts of defending the Town were given up. The Soldiers are gone on board two Transports and those who have dared to be active in supporting Government are under the necessity also of taking refuge in vessels. Such as had not that in their power are left to the mercy of the Rebels who have taken possession of the Town – a single regiment a few weeks*

---

[13] D.R. Anderson, ed., "Colonel Howe to the President of the Virginia Convention, 15 December, 1775," *Richmond College Historical Papers*, 1905, 131

ago would have reduced this colony to a sense of its duty. God only knows when it will be done, now....[14]

Dunmore lamented the panic of his supporters and described the bleak scene in the harbor to Lord Dartmouth:

> *All who were friends of Government took refuge on board of the Ships, with their whole families, and their most valuable Effects, some in the Men of War, some in their own Vessels, others have chartered such as were here, so that our Fleet is at present Numerous tho' not very powerful. I do assure your Lordship it is a most melancholy sight to see the Numbers of Gentlemen of very large property with their Ladies and whole families obliged to betake themselves on board of Ships, at the Season of the year, hardly with the common necessarys of Life, and great numbers of poor people without even these, who must have perished had I not been able to supply them with some flour, which I purchased from His Majesty's service some time ago....*[15]

Despite his precarious situation, Dunmore and his supporters remained safe for the moment, protected in Norfolk harbor by the guns of the British navy, specifically, the 36 gun *Liverpool*, 18 gun *King Fisher*, 14 gun *Otter*, a sloop with 8 guns, the *Dunmore* (formerly Eilbeck, number of guns unknown) and 6 or 7 tenders armed with a few 3 and 4 pound cannon and swivel guns.[16]

---

[14] Clark, ed., "Thomas Macknight to Reverend Macknight, 26 December, 1775," *Naval Documents of the American Revolution*, Vol. 3, 260-61
[15] Ibid. 142
[16] Clark, ed., "Ships in Norfolk and Hampton Roads, 30 December, 1775, *Naval Documents of the American Revolution*, Vol. 3, 309-310

Onshore, the streets of Norfolk swarmed with "rebel" troops. The arrival of Colonel Robert Howe and his North Carolina troops swelled the "rebel" force to over a thousand men. While most encamped out of range of the British naval cannon, guard detachments were posted along the shore to observe Dunmore's activities and warn of a possible attack. Some of the sentries succumbed to the temptation to take pot shots at Dunmore's ships, especially the *Otter*, and frequent flags of truce went back and forth between the two sides concerning the issue of whether the sporadic gunfire from shore was authorized:

Norfolk Dec. 15, 1775
Captain Squire's compliments to the commanding officer, informs him that several musket balls were last night fired at the king's ship from some people at Norfolk. Captain Squire did not return the fire, from a supposition it was done out of wantonness. Captain Squire does not mean to fire on the town of Norfolk unless first fired at; must beg to know if any hostile intention was meant to his Otter sloop....

The Virginia Officers' Reply
Colonel Howe's and colonel Woodford's compliments to captain Squire, and assure him they gave no orders to fire upon the Otter, and conceive the musket balls mentioned in captain Squire's message to have come from our guard, who fired by mistake upon one of our own parties.[17]

---

[17] Clark, ed., "Captain Matthew Squire R.N., to the Officer Commanding at Norfolk, 15 December, 1775," *Naval Documents of the American Revolution*, Vol. 3, 119

Although the sporadic firing continued, it appeared that the situation had reached a stalemate, and Colonel Woodford took the opportunity to request a leave of absence to visit his family.[18] Events soon developed, however, that prevented Woodford's departure.

After two more weeks of indiscriminate gunfire from shore, the proud commanders of the British warships in Norfolk harbor had had enough. Captain Henry Bellew of the H.M.S. *Liverpool* issued an ultimatum:

Captain Bellew to Colonel Howe, Dec. 30, 1775

> *As I hold it incompatible with the Honor of my Commission to suffer Men in Arms against their Sovereign and the Laws, to appear before His Majesty's Ships I desire you will cause your Centinels in the Town of Norfolk to avoid being seen, that Women and Children may not feel the effects of their Audacity, and it would not be imprudent if both were to leave the Town.*[19]

Captain Bellew's request, and his threat, was clear. Rebel sentinels onshore had chided, insulted, and fired upon the ships in the harbor long enough. Either the harassment stopped, or Norfolk would be bombarded.

---

[18] D.R. Anderson, ed., "Colonel Woodford to the President of the Virginia Convention, 15 December, 1775," *Richmond College Historical Papers*, 1905, 130

[19] Clark, ed., "Captain Henry Bellow to Colonel Robert Howe, 30 December, 1775," *Naval Documents of the American Revolution*, Vol. 3, 310

Colonel Howe replied immediately to Captain Bellew's ultimatum with an assurance that the sentinels had been instructed to avoid any insulting behavior and if any were guilty of such conduct, he agreed that they should be punished for it. But, continued Colonel Howe, *"if...you feel it your duty to make your resentment extend farther than merely as to them, we should wish that the Inhabitants of this Town, who have nothing to do in this matter, may have time to remove with their Effects which to Night they have not...."*[20]

The next day -- the last of 1775 -- passed incident free, but on New Year's Day rebel sentinels paraded before the harbor with their hats on their bayonets, taunting the British.[21] Captain Bellew, aboard the *Liverpool*, responded:

> *On the 1ˢᵗ Janry at 3 o'Clock in the Afternoon...their Centinels, came to the Wharf very near me, from their Guard House, which was close to it, and used every mark of insult; I then ordered three Guns to be fired into the House, it had its effect by setting them running; my Lord Dunmore sent (under those Guns) his Boats Arm'd to fetch off a Long boat they had taken from him, whose People set fire to some Store houses, which burnt a good number of Houses, the Rebels have since destroyed the greatest part of the Town.*[22]

---

[20] Clark, ed., "Colonel Robert Howe to Captain Henry Bellew, 30 December, 1775," *Naval Documents of the American Revolution*, Vol. 3, 315

[21] Clark, ed., "A Letter from a Midshipman aboard the Liverpool, 4 January, 1776," *Naval Documents of the American Revolution*, Vol. 3, 621-22

[22] Clark, ed., "Captain Bellew to Philip Stephens, 11 January, 1776," *Naval Documents of the American Revolution*, Vol. 3, 737

Lord Dunmore provided a similar account:

> *Captain Bellow discovering the Rebels parading in the Streets, sent a few Cannon Shot amongst them, his example was immediately followed by all of us, and under Cover of the Cannon, I sent some boats on Shore to burn some detached Warehouses on the lower part of the Wharfs (from whence they used to annoy our boats as they passed) I, at the same time hailed Captain Bellew to beg he would send his boats to burn the Brig with the Salt, which he did immediately.*[23]

Colonel Howe's account of the bombardment was similar to his adversaries, except he placed the blame for the destruction of the city upon the British:

> *The cannonade of the town began about a quarter after three yesterday, from upwards of one hundred pieces of cannon, and continued till near ten at night, without intermission; it then abated a little, and continued till two this morning. Under cover of their guns they landed and set fire to the town in several places near the water, though our men strove to prevent them all in their power; but the houses near the water being chiefly of wood, they took fire immediately, and the fire spread with amazing rapidity. It is now become general, and the whole town will, I doubt not, be consumed in a day or*

---

[23] Clark, ed., "Lord Dunmore to Lord Dartmouth, Aboard the Dunmore off Norfolk, 4 January, 1776," *Naval Documents of the American Revolution*, Vol. 3, 617-18

two.... The burning of the town has made several avenues, which yesterday they had [not], so that they now may fire with greater effect. The tide is now rising and we expect at high water another cannonade. I have only to wish it may be as ineffectual as the last; for we have not one man killed, and but a few wounded. I cannot enter into the melancholy consideration of the women and children running through a crowd of shot to get out of the town, some of them with children at their breasts, a few have, I hear, been killed.[24]

In Colonel Howe's account, the fire that destroyed Norfolk was set, "*in several places near the water*," and spread rapidly because of all of the wooden structures. A number of other accounts from both sides of the engagement suggest that the fires set by Dunmore's men quickly spread throughout the city, but Lord Dunmore claimed that the wind, (which blew towards the water) prevented the fires set along the shore from spreading to the rest of the city and placed the blame for the destruction of Norfolk upon the rebels:

The Vessels from the Fleet to shew their Zeal for His Majesty's Service, sent great Numbers of Boats on Shore, by which means the fire soon became general on the Wharfs, the wind rather blowing off Shore would have prevented the fire from reaching any farther than the Wharfs, but the Rebels so soon as the Men of War ceased firing, and our People came off,

---

[24] Clark, ed., "Colonel Howe to the Virginia Convention, 2 January, 1776," *Naval Documents of the American Revolution*, Vol. 3, 579-80

> *put the finishing Stroke to it, by Setting fire to every House, which has given them employment for these two days past, they have also burnt many houses on both sides of the River, the property of individuals who have never taken any part in this contest, in Short from every transaction they appear to me to have nothing more at heart, than the utter destruction of this once most flourishing Country, Conscious I suppose that they cannot long enjoy it themselves, they wish to make it of as little use as possible to others;*[25]

Dunmore's claim that the "rebels" actually burned the bulk of Norfolk was dismissed by most Virginians as a lie, but an official inquiry of the incident in 1777 (which was not made public for over sixty years) concluded that Dunmore's claim was correct:

> *Upon an inspection of the schedule, and the depositions which have been taken, it will appear that very few of the houses were destroyed by the enemy, either from their cannonade or by the parties they landed on the wharves; indeed, the efforts of these latter were so feeble that we are induced to believe most of the houses which they did set fire to might have been saved had a disposition of that kind prevailed among the soldiery, but they* [the troops under Colonel Howe and Colonel Woodford] *appear*

---

[25] Clark, ed., "Lord Dunmore to Lord Dartmouth, Aboard the Dunmore off Norfolk, 4 January, 1776," *Naval Documents of the American Revolution*, Vol. 3, 617-18

> *to have had no such intentions; on the contrary, they most wantonly set fire to the greater part of the houses within the town, where the enemy never attempted to approach, and where it would have been impossible for them to have penetrated.*[26]

Months of anger at the residents of Norfolk for their cooperation with and in many cases outright support for Dunmore erupted into a combustible fury, and although no one admitted it publicly, Norfolk was looted and burned to the ground by the soldiers under Colonel Howe and Woodford, some of who shouted, "*Keep up the Jigg! Keep up the Jigg!*" as they torched building after building.[27] Nearly 900 structures were destroyed over three days and the destruction spread across the river to Andrew Sprowles shipyard and warehouses in Gosport.[28] Colonel Woodford confirmed the extent of damage, reporting that nine tenths of Norfolk was destroyed.[29]

Unfortunately for Lord Dunmore, the accounts of the devastation that appeared in the gazettes laid the blame upon the governor and his naval lackeys:

> *"It was a shocking scene to see the poor women and children, running about through the fire, and exposed to the guns from the ships, and some of them with children at their breasts. Let our countrymen view*

---

[26] Journal of the House of Delegates, 1835-36, Doc. No. 43, Richmond, 1835, Virginia State Library, 16
[27] Scribner and Tarter, *Revolutionary Virginia: Road to Independence*, Vol. 5, 16
[28] Ibid.
[29] Clark, ed., "Colonel William Woodford to Thomas Elliot, 4 January, 1776," *Naval Documents of the American Revolution*, Vol. 3, 617

and contemplate the scene!...The cannonade had lasted twenty five hours when the express came away, and the flames were raging (it being impossible to extinguish them on account of the heavy fire from the ships) and had consumed two thirds of the town.... It is affirmed that one hundred cannon played on the town almost incessantly for twenty five hours....[30]

Another account from an observer with the "rebels" painted a desolate picture of Norfolk:

I doubt not you have heard before this of the furious cannonade with which the enemy opened the year. They began about 3 in the afternoon of new year's day, and continued, with very little intermission, for 9 hours. Everything that could carry a gun, from a frigate to a boat, played against us. Under the cover of their cannon, they set fire to the town in four different places, and made several attempts to land field pieces, but were repulsed with loss. The horror of the night exceeds description, and gives fresh occasion to lament the consequences of civil war.

The thunder of artillery, the crash of falling houses, the roar of devouring flames, added to the piteous moans and piercing shrieks, of the few remaining wretched, ruined inhabitants, form the outlines of a picture too distressing to behold without a tear. I pray God I may never see the like again.... In short, desolation and ruin have overspread the face of the

---

[30] Clark, ed., "Pinkney Virginia Gazette, Account of the Burning of Norfolk, 6 January, 1776," *Naval Documents of the American Revolution*, Vol. 3, 661

country, and the once populous town of Norfolk now resembles, in miniature the ruins of Palmyra![31]

The destruction of Norfolk did not mean an end to the fighting. The combatants remained where they were, Lord Dunmore's force onboard ships just offshore from the smoldering ruins of Norfolk, and Colonel Howe's troops on the outskirts and amongst the ruins of the city. Heated skirmishes occasionally erupted whenever Dunmore sent landing parties ashore and both sides suffered casualties, but for most of January the two sides shared the common misery of a winter encampment -- wet, cold weather, and limited provisions and supplies.

Norfolk is Abandoned

It was no secret that Colonel Howe and Colonel Woodford wanted to abandon Norfolk and in doing so, burn what was left of the city to deny its use to Dunmore. Both officers argued (before and after Norfolk was burned) that it was dangerous for their force to remain at Norfolk; they ran the risk of being cut off by British reinforcements.[32] Colonel Howe went to Williamsburg in mid-January to update the Convention and gain their approval to withdraw from Norfolk. On January 15th, the Virginia Convention relented and Norfolk

---

[31] Clark, ed., "Extract of a Letter from a Gentleman at Norfolk, 7 January, 1776," *Naval Documents of the American Revolution*, Vol. 3, 673 Printed in Pinkney's Virginia Gazette, 20 January, 1776

[32] Anderson, ed., "Colonel Robert Howe to Edmund Pendleton, 22 December, 1775," and 2 January, 1776, and Colonel William Woodford to Edmund Pendleton, 22 December, 1775," *Richmond College Historical Papers*, 136-39, 148

was abandoned in early February.[33] The bulk of the rebel troops took up a new post in Suffolk, twenty miles southwest of the ruins at Norfolk. Additional detachments were posted at Kemp's Landing and Great Bridge to block provisions from reaching Dunmore.[34] Before the troops left, they torched the remaining buildings of Norfolk, over 400 of them.[35]

Within a week of Howe's departure, Lord Dunmore, covered by the 44 gun frigate *H.M.S. Roebuck* (which sailed into Norfolk harbor three days after the "shirtmen" marched to Suffolk), landed troops on Tucker's Point (adjacent to Portsmouth and across the river from Norfolk). They immediately dug wells to replenish their critically low supply of fresh water. A windmill and a few buildings stood on the point, damaged, but not destroyed in all of the fighting, and Dunmore converted some of the buildings into barracks to house his growing number of smallpox cases.[36] Earthworks were erected to, *"Secure the Watering Place from the Depredations of the Rebels,"* and ovens were built to supply bread.[37] It was clear that Lord Dunmore, the most despised man in Virginia, had no intention of leaving the colony.

Prior to the destruction of Norfolk and the withdrawal of American forces to Suffolk, Colonel Woodford requested

---

[33] Scribner and Tarter, ed., "Proceedings of the 4th Virginia Convention, 15 January, 1776," *Revolutionary Virginia, Road to Independence,* Vol. 5, 405

[34] Clark, ed., "Purdie Virginia Gazette, 9 February, 1776," and "Letter in London Chronicle,," *Naval Documents of the American Revolution,* Vol. 3, 1187   and   Vol. 4, 23

[35] Journal of the House of Delegates, 1835-36, Doc. No. 43, Richmond, 1835, Virginia State Library, 16

[36] John Selby, *The Revolution in Virginia, 1775-1783,* (Colonial Williamsburg Foundation, 1988), 86

[37] Clark, ed., "Journal of HMS Liverpool, 13-14 February, 1776," *Naval Documents of the American Revolution,* Vol. 3, 1293   and Selby, 86

permission from the Virginia Convention to return home to Caroline County. The weeks following his success at Great Bridge had been difficult on the tired commander of the 2nd Virginia Regiment. Although he never admitted it, he was likely a bit annoyed at having to surrender command to Colonel Howe after his decisive victory at Great Bridge. His relations with Colonel Henry remained strained and extended to both a detachment of 1st Virginia troops that had joined him from Williamsburg after the Battle of Great Bridge as well as his own troops, who, like most Virginians, revered Patrick Henry.

Chapter Six

"I Must Request Your Permission to Retire."

1776

One incident in particular greatly disturbed Colonel Woodford because he felt it undermined both his authority and overall military discipline. In late December, a court-martial refused to convict Lieutenant Francis Boykin of Captain ---- Davis's company of the 1st Virginia Regiment of any wrongdoing following Colonel Woodford's arrest of him for abandoning his post (with a detachment of men). According to testimony at Boykin's court martial, Lieutenant Boykin was ordered to guard Batchelors Mill, about mid-way between Portsmouth and Great Bridge on Deep Creek, a tributary of the South Branch of the Elizabeth River.[1] Boykin claimed that he and his 25 man detachment were essentially abandoned there and after eleven days of hardship without proper shelter, provisions, or any hint of relief, Lieutenant Boykin led his men back to Norfolk.[2]

Colonel Woodford confronted Boykin upon his arrival and asked if Boykin had any orders to quit his post, to which Boykin replied, "*He had rather be broke than suffer again as

---

[1] Anderson, ed., "Court-Martial of Lieutenant Francis Boykin, 26 December, 1775," *Richmond College Historical Papers*, 141-143
[2] Ibid.

much as he did at the Mill."[3] Colonel Woodford accommodated Boykin and arrested him for deserting his post.

At his court martial, Boykin testified that he and his men had endured several days of half rations of meat and that his men, contrary to his orders, had disposed of all of the local poultry to supplement their provisions.[4] Woodford noted that the detachment had plenty of flour and bread, but Boykin's assertion that he and his men went eleven days without any relief, during which, *"neither himself nor his men had a house to lie in & himself and many of them not a blanket or Rug to cover them,"* convinced the officers of the court martial that Boykin's actions were justified.[5]

Colonel Woodford was furious and appealed the tribunal's decision to the Virginia Convention, noting the danger to military discipline should Boykin go unpunished. In his appeal, Woodford argued that if Boykin had actually been as distressed as he claimed, he could have marched to Great Bridge or Kemp's Landing (two posts he marched through on his way to Norfolk) and obtained the provisions he lacked, then returned to his post at Bachelor's Mill. Woodford also insisted that Boykin's acquittal for, *"a crime that he conceives to be one of the greatest an officer can be guilty of, will have a very bad tendency & be a most alarming example to other Officers as well as the private men who were led from their duty by the person the Colony depended upon to Keep them to it."*[6]

---

[3] Ibid.
[4] Ibid.
[5] Ibid.
[6] Anderson, ed., "Colonel William Woodford to Colonel Robert Howe, 27 December, 1775," *Richmond College Historical Papers,* 14

Although the Convention appeared sympathetic to Woodford's appeal and even censured Lieutenant Boykin, they acknowledged that they had no authority to overrule the decision of the military court in this matter.[7]

Annoyed by Woodford's appeal to the Convention, Lieutenant Boykin mocked the colonel in the newspapers, writing an open letter questioning by what authority Woodford had appealed the verdict.[8] Boykin added that Woodford's actions towards him were likely the reason so many officers in Norfolk were discontented with Woodford's command.[9]

This insinuation prompted yet another public letter, this one from Major Richard Kidder Meade, who asserted that

> *On reading Lieutenant Boykin's queries, I saw no room for the public to suppose but that a discontent prevailed in the breast of every officer here* [in Norfolk]. *As he made no exceptions, I think proper to declare, for myself and officers, that no such discontent has been amongst us....*[10]

Although Colonel Woodford undoubtedly appreciated Major Meade's expression of support, he was likely stung by the whole affair and received some sympathy from his friend, and President of the Committee of Safety, Edmund Pendleton:

> *I cannot avoid feeling deeply for your disagreeable situation – confined in a dirty place – harassed with*

---

[7] Mays, ed., "Edmund Pendleton to Colonel Woodford, 5 January, 1776," *The Letters and Papers of Edmund Pendleton*, Vol. 1, 147

[8] Dixon and Hunter, *Virginia Gazette*, 13 January, 1776, 3

[9] Ibid.

[10] Purdie, *Virginia Gazette*, 2 February, 1776, 3

variety of duty – and chagrined by a popular opposition from inferior officers.[11]

Pendleton assured Woodford though, that the colonel maintained the support of the Convention. *"You were commended, not blamed for making* [the appeal]*"*.[12] Pendleton also expressed concern for Woodford's health, noting the feebleness of Woodford's hand in his last letter, and informed Colonel Woodford that he would endeavor that morning to procure Woodford's desired leave of absence.[13] Three more weeks passed, however, before Colonel Woodford was able to leave camp and head home to Caroline County.

Woodford Rests at Windsor

Colonel Woodford was undoubtedly relieved to return to his wife and two sons at Windsor in Caroline County. Letters from Edmund Pendleton suggest that Woodford needed the rest to recover his health. We know little about how else he spent his two month leave of absence. It is likely that much of his time was devoted to managing his estate as well as his brewery interests in Fredericksburg.

Several developments occurred during Woodford's leave of absence that significantly altered the structure of Virginia's military forces. In February, word arrived that the Continental Congress in Philadelphia had taken authority of Virginia's regiments of regulars (increased now to eight by the Virginia Convention) and had appointed Andrew Lewis, the

---

[11] Mays, ed., "Edmund Pendleton to Colonel Woodford, 5 January, 1776," *The Letters and Papers of Edmund Pendleton*, Vol. 1, 147
[12] Ibid.
[13] Ibid.

experienced militia commander of Lord Dunmore's expedition against the Shawnee in 1774, as Brigadier General of Virginia's continental forces. This made both Colonel Woodford and Colonel Henry subordinate to Lewis, a development that troubled both commanders. This blow to Patrick Henry's pride, which he believed occurred with the blessing of the Committee of Safety, was too much for him to accept, and he resigned his commission.

The news of Henry's resignation prompted a near mutiny in the ranks of the 1st Virginia; they went into, *"deep mourning"* and gathered, under arms at Henry's lodging in Williamsburg to address him.[14] Henry graciously thanked those assembled for their support and then attended a farewell dinner at the Raleigh Tavern in his honor. He was forced to postpone his departure from the capital, however, when word spread of, *"some uneasiness getting among the soldiery, who assembled in a tumultuous manner, and demanded their discharge, declaring their unwillingness to serve under any other commander."*[15] Henry spent most of the evening with the troops, *"visiting the several barracks, and* [using] *every argument in his power with the soldiery to lay aside their imprudent resolution, and continue in the service..."*[16] His efforts succeeded and the disgruntled troops eventually settled down.

---

[14] Purdie, *Virginia Gazette*, 1 March, 1776, 3
[15] Ibid.
[16] Ibid.

General Charles Lee Arrives

The arrival of General Charles Lee in Williamsburg in late March offered a solution to the discipline dilemma that had long plagued the Virginia regiments. Lee, a former British officer with extensive military service in Europe, held the rank of Major-General in the Continental army and served with General Washington in Massachusetts in 1775. Although he was a native of Britain and had only arrived in the colonies in 1773, Lee had earned the trust and admiration of many in Congress and held the third highest rank in the continental army. He was the most militarily experienced and knowledgeable officer in the army and was highly esteemed throughout the colonies.

Reports of planned British military operations in the southern colonies prompted Congress to send General Lee southward in March to oversee the region's defense. He arrived in Williamsburg on March 29th and wrote to General Washington about the situation he found:

> *The Regiments in general are very compleat in numbers, the Men (those that I have seen) fine – but a most horrid deficiency of Arms – no entrenching tools, no* [effective cannon] *(although the Province is pretty well stockd)...I have order'd...the Artificers to work night and day....*[17]

Lee speculated that the unrealistic hope for reconciliation with Britain among some Virginians and a degree of apathy among

---

[17] Philander D. Chase, ed., "General Charles Lee to General George Washington, 5 April, 1776," *The Papers of George Washington*, Revolutionary War Series, Vol. 4, (Charlottesville, VA: University of Virginia Press, 1991), 43

others had caused Virginia to procrastinate on important military preparations. He also criticized the scattered deployment of the colony's regiments (which had been posted upon the several peninsulas of eastern Virginia to guard against any water borne incursions by Lord Dunmore or the British). Lee noted sarcastically that, *"They have distributed their Troops in so ingenious a manner, as to render every active offensive operation impossible."*[18]

General Lee acted quickly to rectify the situation. He ordered the 5th Regiment and half of the 7th Regiment to march from their duty stations in Richmond Courthouse and Gloucester Courthouse to Williamsburg (to join the 1st and 6th regiments who were already in the capital).[19] He commenced work on fortifications for Jamestown, Burwell's Ferry, Yorktown, and Williamsburg and publically urged Virginia's young gentlemen (in the gazettes) to voluntarily form companies of light dragoons, something the convention had failed to do because of the expense.[20] When a shortage of muskets left many of the troops unarmed, Lee resorted to the use of spears, arming two companies of the tallest and strongest men of each regiment with them.[21]

---

[18] Ibid.
[19] Robert Schribner and Brent Tarter, ed. *Revolutionary Virginia, Road to Independence*, Vol. 6, (University Press of Virginia, 1981), 277
[20] Ibid., 278 and Purdie, *Virginia Gazette*, 26 April, 1776, 1
[21] Chase, ed., "General Charles Lee to General George Washington, 10 May, 1776, *The Papers of George Washington*, Vol. 4, 258

Colonel Woodford Returns to the Army

In early April, Colonel Woodford, rested and recovered, rode into Williamsburg and reported to General Lee. Woodford presided over a council of officers on April 6th to consider Lee's controversial proposal to forcibly evacuate a portion of the populace of Norfolk and Princess Ann counties in order to deny Lord Dunmore much needed provisions from the local inhabitants. The council agreed with the need to remove the inhabitants and their livestock out of reach and contact with Dunmore, provided however that, *"proper provision could be made for these unfortunate people after their removal."*[22]

Colonel Woodford remained in Williamsburg for two more weeks and then accompanied General Lee to Suffolk, where he resumed command of his regiment (and all of Virginia's forces south of the James River upon Lee's return to Williamsburg at the end of April).

General Lee superintended the beginning of the forced evacuation of the region, but left Colonel Woodford in charge of completing the task. Although he had supported the policy in the officer's council, Woodford found its implementation extremely disagreeable and expressed his discomfort to his wife:

---

[22] "At a Council of Officers at Head Quarters, Williamsburg, April 6, 1776," *The Lee Papers*, Vol. 1, (Collections of the New York Historical Society, 1871), 387

My Dearest Molly,
> *I wrote you from Suffolk when I informed you I should attend the General from this place* [Kemp's Landing], *but little thought I was to be left behind to execute the most disagreeable pieces of service that could have happened, viz: the removing all the Inhabitants of these two Tory Countys.* [at least 12,000 souls]. *Their distress will be great indeed, and I expect to be witness to many more affecting scenes than even Norfolk afforded.*[23]

Woodford had apparently hoped to return with General Lee to Williamsburg and had written to his wife about joining him there, but those plans were dashed by his new assignment.

> *This (with the additional thought of being deprived of seeing you on the first of May) has perplexed me much, but the service of my country must be the first consideration, and where Duty calls, pleasures and conveniences must give way. I have no doubt but you will summon all your Fortitude upon this disappointment. From the orders left by the Gen. when he returned I see no prospect of my Regiment returning to Williamsburg,* [you will] *therefore have to hold yourself in constant readiness to come down, whenever I send for you. If we should be continued* [here] *I could make Suffolk tolerably agreeable to you. Write me what you think of coming there....*[24]

---

[23] Stewart, "Colonel Woodford to his Wife, April 30, 1776," *The Life of Brigadier General William Woodford of the American Revolution* Vol. 1, 650-651
[24] Ibid.

Woodford included comments on recent purchases he had made for his wife and children and then described a recent visit to Portsmouth with General Lee:

> *I was the other day at Portsmouth with the Gen. to view the Enemy's entrenchments, and visit my old neighbors, they saluted us with one gun from Capt. Squire and a few swivels and small arms from a Tender, which proved as Impotent as usual. All the inhabitants of that Beautiful Village and its neighborhood is removed, the gardens and wells destroyed.*[25]

In a letter to General Lee two days later, Woodford reiterated his discomfort with his assignment and confessed that he had taken some liberty to relieve some of the suffering of the inhabitants:

> *The distress of these unfortunate people is hardly to be described, it's the most disagreeable service I have ever been engaged in, you were pleased to leave several things at my discretion. I have ventured to give numbers of them passes to remove into the interior parts of N. Carolina, and where they could be best accommodated.*[26]

General Lee expressed sympathy for Woodford, writing that, "*I pity you most sincerely for your damn'd employment...*" but Lee reasserted his, and the Virginia Convention's

---

[25] Ibid.
[26] "Col. Wm. Woodford to Gen. Lee, May 2, 1776," *The Lee Papers*, Vol. 1, 462

determination to deprive Dunmore the support he had received all winter from the countryside.[27]

At the time General Lee wrote to Woodford, he was making final preparations to march for the Carolinas with the 8th Virginia Regiment to counteract British activity further south. British General Henry Clinton commanded a powerful British force that had actually sailed into Chesapeake Bay in February before Lee even arrived in Virginia. Clinton's stay in Virginia was brief (much to the chagrin of Lord Dunmore) and his force soon set sail for open water, destined ultimately for Charlestown, South Carolina. Although General Lee was unaware of Clinton's specific plans, he was determined to defend the Carolinas and rode south with the 8th Virginia in mid-May to confront Clinton.

Dunmore Departs

It appears that the forced evacuation of the Norfolk and Princess Ann County coastline that Colonel Woodford found so distasteful to implement may have helped persuade Lord Dunmore to abandon his position in southeastern Virginia. On May 22nd, Dunmore's troops left the earthworks defending Tucker's Point and boarded ships offshore. Dunmore then ordered his fleet of approximately 70 vessels to sail down the Elizabeth River. Colonel Woodford informed General Andrew Lewis in Williamsburg (who had assumed command of Virginia's forces in General Lee's absence) of Dunmore's movement. In doing so, Woodford took credit for forcing Dunmore's action:

---

[27] "Gen. Lee to Col. Wm. Woodford, May 11, 1776," *The Lee Papers*, Vol. 2, 23

The vigilance of my guards has occasioned the enemy to abandon their lines at Portsmouth. This, and some fire rafts I was preparing, has likewise occasioned the fleet to go off. They have thrown over their salt, burnt the most indifferent of their small craft, and are all now below Crany island, except four ships, which are opposite the distillery, but under way likewise. One of the 14th regiment, and five sailors, have deserted, they inform me they have the smallpox. I have given very particular orders to avoid this evil, if it be true. They all concur in the same story, and likewise that the fleet is bound for Cape Fear; but I doubt whether my Lord does not intend a secret expedition to some other part of the colony....[28]

Woodford's speculation of Lord Dunmore's destination proved to be well founded. At first it appeared that Dunmore and his fleet were heading out to sea, but instead of passing the Virginia Capes, Dunmore tacked north and sailed up Chesapeake Bay, anchoring in the mouth of the Piankatank River. Dunmore had decided to use Gwynn's Island, just offshore from the Virginia mainland, as his new base of operation.

Woodford Passed Over for Promotion

Lord Dunmore's surprising move caused General Lewis to order Colonel Woodford and the 2nd Virginia Regiment to Williamsburg to help defend the capital. Woodford

---

[28] Dixon and Hunter, "Extract of a letter from Col. Woodford to General Lewis, dated Norfolk, May 22," *Virginia Gazette*, 25 May, 1776, 3

undoubtedly welcomed the order and arrived in Williamsburg in the end of May. He found the city abuzz with political and military activity. The 5th Virginia Convention had already taken the decisive step (unanimously) of instructing Virginia's delegates to the 2nd Continental Congress to propose independence for all of the colonies and were now engaged in forming a new constitution. Woodford's friend, Edmund Pendleton, played a leading role in both actions.

On the military front, the situation at Gwynn's Island had settled into an uneasy stalemate with the Virginians unable to strike Dunmore on the island and Dunmore unwilling to risk battle on the mainland. Colonel Woodford attended to both his regiment and General Lewis, but was stunned in mid-June to learn that Colonel Hugh Mercer of the 3rd Virginia Regiment had been promoted ahead of Woodford to the rank of Brigadier-General in the continental army.

Woodford admired and respected Mercer and had even declared on several occasions a willingness to serve under his command. The experienced Scottish officer had held the rank of colonel during the French and Indian War and had commanded Pennsylvania troops during that long conflict. Woodford's declarations of support for Mercer, however, were months old and Woodford had assumed that when Congress had taken Virginia's regiments into continental service back in February, the issue of rank had been settled with him holding seniority over Mercer due to his earlier appointment.

Colonel Woodford felt that Mercer's promotion ahead of him would be viewed as an aspersion on his character, and to make matters worse, rumors that yet another officer, Colonel Adam Stephens of the 4th Virginia Regiment, was also to be promoted ahead of Woodford, prompted Colonel Woodford to write to General Washington to express his displeasure and offer his resignation:

> *I am sorry to trouble you with complaints, but give me leave Sir to say I feel myself much hurt by the late promotion of my very worthy friend Col. Mercer, and to request your patience to hear my reasons in the best manner I am capable of giving them, with that freedom, which I flatter myself will not be taken amiss by you.*
>
> *When the military establishment of this colony first took place, I offered my service in any post the convention thought proper to appoint me to, without soliciting any one man of that Body for his vote or interest before the ballot began, I informed the House* [3rd Virginia Convention] *of which I was at that time a Member that I wished to serve under that Gentleman, and desired no person would vote for me in preference to him. Notwithstanding all that could be said he was rejected, and my appointment confirmed. When the Honorable, the Congress, took* [our] *troops upon the Continental Establishment, a few months ago, I again expressed my wish that Colonel Mercer might be appointed to a higher officer, their wisdom directed them to make the arrangement otherways, and I looked upon the army as firmly established in such a manner, that every officer would rise in his turn, unless some fault could be laid to his charge. I have the same good opinion of that gentleman I ever had, but what I complain of*

> *is the impropriety as I conceive of the appointment, and that the promotion of an officer at that time serving under me, (however well he may deserve it) reflected dishonor upon myself, and will be attributed by the world to some misconduct in me, or at best inability to fill a higher office. I am informed from good authority, that a similar promotion is now in contemplation in favor of Col. Stephens. From the above reasons, I must request your permission to retire, not with any intention to promote any disturbances either, in the army, or country, but on the contrary to do any future service in a private way, to my country, and the common cause, to which I feel myself as warmly attached as ever.[29]*

Woodford hinted that he might be persuaded to stay in the army if his grievance might somehow be addressed.

> *Before I conclude I will take the liberty to appeal to your own feelings as an officer, upon such an occasion, and to ask you what light I must be looked upon in the army for the future. My opposition to a popular military Officer* [Patrick Henry] *and my exertion to introduce some discipline among those infant troops, has gained me enemys, who I can see exulting in the late promotion, though they hate the man* [Mercer].[30]

---

[29] Philander D. Chase, ed., "Colonel William Woodford to General Washington, July 6, 1776," *The Papers of George Washington*, Vol. 5, (Charlottesville, VA: University Press of Virginia, 1993), 228-230
[30] Ibid.

When Woodford wrote to General Washington on July 6th, the military standoff at Gwynn's Island had entered its sixth week, but General Lewis and the Virginian troops facing Dunmore were about to alter the situation. Lewis, accompanied by Colonel Woodford and several other officers, arrived at Gwynn's Island on July 9th and ordered recently completed artillery batteries erected along the shore to open fire of Lord Dunmore's troops and ships.

Apparently caught by surprise, Dunmore evacuated the island by nightfall, returning to his ships. He lingered in Chesapeake Bay for a few more weeks, but finally gave up efforts to regain control of Virginia and set sail for New York in August. Virginia was now free to send troops north to reinforce its native son, General Washington, and his army in New York.

Following the brief battle at Gwynn's Island, Colonel Woodford returned to Williamsburg to command his regiment. The term of enlistment for his men was about to expire so he took action to re-enlist as many as he could and recruit new troops. Woodford's efforts to instill more discipline in the troops over the course of the past year, however, hampered his efforts. Many of his men refused to re-enlist under his command and it soon became apparent that the 2nd Virginia Regiment was in no condition to march north to join General Washington in New York. An observer in Williamsburg noted that the efforts of both General Lewis and Colonel Woodford failed to convince the soldiers of the 2nd Regiment to re-enlist:

> In obedience to Congress, two Regiments are ordered to N. York instantly. Gen'l Lewis, as a lure to the 1st and 2nd, directed that they should be re-enlisted for 3 years to seize the post of Honour as he terms it, hoping that the men's well grounded Complaints would thus be hushed into peace. But alas!, human nature is not so easily smothered and to Col. Woodford's great mortification, the 1st almost to a man swallowed the bait, while his 2nd resisted his [Woodford's] eloquent harangue at their head, and silently rejected the intended honour he proposed doing them by delaying his resignation that he might lead them on the Field of Glory. They say they will follow Col. Scott, but he is ordered to the 5th and I question much whether Col. [Woodford] will immediately resign, tho he is certain they will re-enlist; twill be tried tomorrow.[31]

Woodford's men were not the only ones unwilling to extend their service, Colonel Woodford himself had made inquiries of his friend Edmund Pendleton about the possibility of representing Caroline County in the new House of Delegates after he resigned his commission. Pendleton explained to Woodford that his chances to gain a seat that summer were poor and that he might have to wait until spring for a better chance to serve in the government.

> Should you therefore leave the Army," wrote Pendleton on July 31st, *You must be content to wait*

---

[31] Stewart, "Gabriel Johnston to Leven Powell Aug. 6, 1776," *The Life of Brigadier General William Woodford of the American Revolution* Vol. 1, 705

'til next Spring, when New Elections will make room for you. Pendleton added, *"I shall be happy to see you in the Assembly, at the same time I fear bad consequences from your leaving the Army."*[32]

Woodford's low popularity among his troops was not improved by his response to a letter published in the newspapers from the members of the 1st and 2nd Virginia Regiments congratulating Patrick Henry for his election as governor of Virginia in late June:

> *Permit us,* [declared the letter on July 5th] *with the sincerest sentiments of respect and joy, to congratulate your Excellency upon your unsolicited promotion to the highest honours a grateful people can bestow.*
>
> *Uninfluenced by private ambition, regardless of sordid interest, you have uniformly pursued the general good of your country; and have taught the world, that an ingenuous love of the rights of mankind, an inflexible resolution, and a steady perseverance in the practice of every private and publick virtue, lend directly to preferment, and give the best title to the honours of an uncorrupted and vigorous state.*
>
> *Once happy under your military command, we hope for more extensive blessings from your civil administration....*[33]

---

[32] Mays, ed., "William Woodford to Edmund Pendleton, July 31, 1776," *The Letters and Papers of Edmund Pendleton*, Vol. 1, 190

[33] Purdie, "To His Excellency Patrick Henry, jun. esq., Governour of the Commonwealth of Virginia," *Virginia Gazette*, July 5, 1776, 3

Colonel Woodford was apparently furious that the wording of this letter suggested that the colonel of the 2nd Virginia Regiment agreed with the letter's sentiments and he demanded from his officers that this mistaken view be cleared up. As a result, the officers submitted a second public letter to Purdie's gazette in early August.

> *Mr. Purdie,*
> *Let the publick know that col. Woodford's name was not among the subscribers of the address to the Governour; that it was not presented as containing the sentiments of the colonel, but of the officers and their men, and that the col. was not consulted on the occasion. This piece of justice is demanded by the colonel, and cheerfully granted by the officers.*[34]

Woodford Resigns His Commission in Protest

Despite the numerous frustrations and disappointments of the summer, Colonel Woodford remained with the army until early September, when confirmation of Adam Stephen's promotion to brigadier-general finally arrived. That was the final straw for Woodford; he resigned his commission in protest and returned to Caroline County.

Edmund Pendleton wrote to console and advise his friend the following month:

> *Tho' I must lament the Occasion which has deprived Our Cause of so promising an Officer, yet I cannot disapprove of your resolution to resign a Post wherein your Honor has been wounded; I will not*

---
[34] Purdie, *Virginia Gazette*, Aug. 9, 1776, 3

assert it, but am not without my Suspicions, that what has happened to you, was part of the business settled in a certain Cabinet our last Convention, since I am told by Colonel Harrison that it seemed agreed when he left Congress that you and Stephen should be promoted together the next appointment which was made, and yet they were picking up any thing North or South to shun you.[35]

Woodford had apparently informed Pendleton after his resignation that he intended to travel to New York to meet with General Washington, for Pendleton added, "*I cannot but approve your Resolution of visiting the Camp, as it will silence Insinuations which some people might make, however groundless, that the want of Courage had produced your Resignation at this critical time.*"[36]

Colonel Woodford rode north in early November and delivered a letter to General Washington from Washington's brother, John Augustine, of Westmoreland County.[37] Woodford found the American army in disarray, chased out of New York and fleeing across New Jersey. As a private citizen there was little Woodford could do to help and his stay in camp was brief.

While he most likely met with General Washington to deliver the letter from John Augustine, we know nothing of what was discussed and Woodford departed for Virginia

---

[35] Mays, ed., "Edmund Pendleton to William Woodford, Oct. 11, 1776," *The Letters and Papers of Edmund Pendleton*, Vol. 1, 202-203

[36] Ibid.

[37] John C. Fitzgerald, "General Washington to John Augustine Washington, November 19, 1776," Writings of George Washington, Vol. 6, (Washington, DC: U.S. Govt. Printing Officer, 1932), 245

within days of his arrival with General Washington's response to his brother.

The Game is Pretty Near Up

William Woodford must have felt some degree of guilt in his departure from General Washington's camp in November 1776, for the American cause was in jeopardy. Washington's army had largely dissolved in the face of a much larger and superior British and Hessian force and it was doubtful if the remnants of the American army would survive the winter. Congress abandoned Philadelphia in mid-December, expecting an inevitable British assault of the de facto American capital while General Washington and just a few thousand troops guarded the crossing points of the Delaware River around Trenton, New Jersey. Knowing that they had little hope of stopping the British from crossing the river, many of Washington's troops just hoped to survive until the end of the year when about half of the army's enlistments expired and they would be free to return home. The situation confronting General Washington in late 1776 was dire indeed.

Alas, fortune shined upon the Americans when General William Howe, the overall British commander in America, opted to halt active operations in mid-December and go into winter quarters. He posted his troops throughout New Jersey (so that they could properly shelter themselves in towns) and waited until spring to finish the Americans off.

General Washington seized the opportunity presented him to turn the tables on Howe and crossed the Delaware River to smash a Hessian detachment posted at Trenton on December 26th, with an early morning surprise attack. Washington followed up his victory at Trenton on January 3rd, with another stunning victory at Princeton, then marched his victorious troops to Morristown, New Jersey where they encamped for the winter on excellent defensive terrain.

General Howe now realized that he had over-extended his troops and withdrew his army closer to New York, abandoning most of New Jersey back to the American rebels. General Washington's bold actions in late December and early January likely saved the American cause and inspired thousands of men to enlist in the American army in 1777. It remained to be seen whether William Woodford would return to the army.

Chapter Seven

"The Rebels Disputed [the ground] *with Great Spirit, Particularly Their Officers"*

1777

We unfortunately know little of what occurred over the winter with William Woodford following his return to Virginia in late November 1776. He had seen firsthand the dire condition of the American army, in disarray and retreat, and when he left to return to Virginia, he likely expected that Philadelphia would soon join New York as an occupied city of the British. Stunning and unexpected American victories at Trenton and Princeton, however, saved Philadelphia and revived the flagging spirits of the patriot cause. The joy William Woodford felt at General Washington's dramatic victories in New Jersey was likely offset by the sad news of General Hugh Mercer's death from wounds he received at Princeton. Woodford held Mercer in high regard, and his loss was a significant blow to the American army.

In early March, 1777 Woodford received news he had long hoped for. Despite his earlier resignation, the Continental Congress had appointed him a brigadier-general. It was initially uncertain, however, whether Woodford would accept the commission, for he was ranked as the last of eleven newly appointed brigadier-generals in the American army.

Woodford's allies in Congress argued that his experience and previous service should place Woodford near the top of the list of new brigadier-generals, senior to fellow Virginians (and former subordinate officers) George Weedon and Peter Muhlenberg, but the majority of Congress, citing Woodford's resignation from the army, placed him at the bottom of the list.[1]

Concern about Woodford's placement among the generals (and how Woodford would take the news) extended to General Washington, who wrote a candid letter to Woodford upon learning of his appointment to urge him to accept the commission:

> *By some Resolves of Congress, just come to my hands, I find as I hoped and expected, your name in the new appointment of Brigadiers; but perceivd at the same time, that you were named after Muhlenberg and Weedon – the reason assign'd for this – your having resign'd your former Rank in the Service of the Continent.*
>
> *You may well recollect my dear Sir, that I strongly advised you against this resignation – I now as strongly recommend your acceptance of the present Appointment – You may feel somewhat hurt, in having two Officers placed before you (tho' perhaps never to command you) who once were inferior in point of Rank to you; but remember, that this is a consequence of your own Act. And – consider what a stake we are contending for --- Trifling punctilios*

---

[1] Worthington C. Ford, ed., "22 February, 1777," *Journals of the Continental Congress*, Vol. 7, 141-142

should have no Influence upon a Mans conduct in such a Cause; and at Such a time as this – If Smaller matters do not yield to greater; If trifles, light as Air; in comparison of what we are contending for; can withdraw, or withhold Gentlemen from Service, when our All is at Stake, and a single cast of the Die may turn the Tables, what are we to expect – It is not a common Contest we are Ingaged In – every thing valuable to us depends upon the Success of it – and the success upon a speedy, & vigorous Exertion. Consider twice, therefore, before you refuse.[2]

Woodford's friend, Edmund Pendleton, recognized the awkward position Woodford was in and expressed his concern to Richard Henry Lee, a delegate in the Congress:

I observe by the last papers Colonel Woodford is at last promoted and felt concern at seeing him behind Muhlenburg and Weedon; Mercer and Stevens had originally a Right to command him and it was owing to some untoward circumstances, contrary to his endeavor, that he was put over them, and therefore it was just they should be put into their proper places tho' it must hurt the delicacy of a good Officer to have a man under him to-day command him tomorrow; But these Gentlemen, however worthy I think them, had no such claim and I am persuaded would have been happy in Ranking under him. What he will determine to do, I know not, but as I think him a valuable Officer, I wish

---

[2] Frank E. Grizzard, Jr., ed., "General Washington to Brigadier General William Woodford, 3 March, 1777," *The Papers of George Washington*, Vol. 8, (Charlottesville, VA: University of VA Press, 1998), 507-508

for the common good, he may wave all these considerations and return into the service.[3]

After several weeks of reflection, Woodford took Washington's advice to heart and accepted his appointment, arriving in the American camp, which was spread among the Watchtung Mountains in New Jersey from Morristown to Bound Brook, in early May. He found an army on the rebound, growing in numbers and confidence.

Virginia's continental regiments had increased to sixteen in number by 1777 and most of them were in camp (albeit undermanned) when General Woodford arrived in Morristown. Woodford was given command of a brigade made up of the 3rd, 7th, 11th, and 15th Virginia Regiments. At full strength Woodford's brigade would have numbered approximately 3,000 men, but the 15th regiment had yet to arrive and the other three regiments totaled just 999 men on May 20th.[4] Just three days earlier, a return of the twelve Virginia Battalions in Morristown (which included the 3rd, 7th, and 11th Virginia Regiments) reported General Woodfood's effective strength (men fit for duty) at only 555 men, so the increase over three days likely included soldiers who were present but not fit for duty or perhaps detached on assignment.[5] Whatever the cause for the discrepancy,

---

[3] Mayes, ed., "Edmund Pendleton to Richard Henry Lee, 9 March, 1777," *The Letters and Papers of Edmund Pendleton*, Vol. 1, 206

[4] Philander D. Chase, ed., Note: The strength of each regiment of Woodford's brigade was reported on May 20, 1777 to be: 3rd VA -- 150; 7th VA -- 472; 11th VA – 377; 15th VA – Not Yet Arrived. "Enclosure 20 May, 1777," *The Papers of George Washington*, Vol. 9, (Charlottesville, VA: University of Virginia Press, 1999), 492

[5] "A General Return of the 12 Virginia Battalions in Morristown, May 17, 1777," *The Papers of George Washington*, Library of Congress online.

Woodford's brigade, like all of General Washington's army, was growing in size just in time for the upcoming campaign season.

General Woodford's 3rd Virginia Brigade, along with a 4th brigade of Virginians commanded by Woodford's former subordinate in the 2nd Virginia Regiment, General Charles Scott, comprised a division of troops commanded by Major General Adam Stephen. It was Stephen's promotion over Woodford in the summer of 1776 that sparked Woodford's resignation so his attachment to Stephen's division must have been a bit awkward for Woodford.

Of more concern to Woodford, however, remained the placement of two former subordinates, Peter Muhlenberg and George Weedon, ahead of him in seniority. Each commanded their own brigade (the 1st and 2nd Virginia Brigades) and both were attached to General Nathanael Greene's division. This arrangement helped reduce friction between Woodford and his former subordinates as it virtually eliminated the possibility that Woodford would ever fall directly under Muhlenberg's or Weedon's command, but Woodford remained troubled by his reduced station. For the time being, however, he set his feelings aside and concentrated on preparing his brigade for the upcoming campaign.

Woodford's Brigade

General Woodford was fortunate to have the most experienced combat unit of Virginians in his brigade, the 3rd Virginia Regiment. Originally commanded by Hugh Mercer and then George Weedon, the 3rd Virginia was under the command of Colonel Thomas Marshall of Fauquier County in the spring of 1777. These seasoned troops had been the first regiment of Virginian continentals to join General Washington in New York in 1776, had seen action at Harlem Heights and White Plains, and had served in the rear guard of the long American retreat to the Delaware River in late 1776. The 3rd Virginia also fought in the crucial American victories at Trenton and Princeton. With just 167 officers and men fit for duty in late May 1777, the regiment was only at a quarter of its authorized strength, but every soldier in the regiment was an experienced veteran and they would soon demonstrate their worth in the field.[6]

The 7th Virginia Regiment, 219 strong in late May, also possessed some combat experience, having forced the last British royal governor of Virginia, Lord Dunmore, off Gwynn's Island and eventually out of Virginia in the summer of 1776.[7] The 7th Virginia was not ordered north to join General Washington until January of 1777 so they missed the entire New York campaign as well as the action at Trenton and Princeton. They arrived in Morristown in the spring after being inoculated for smallpox and were commanded by Colonel Alexander McClanahan of Augusta County.

The 11th Virginia Regiment was perhaps the most unique of all of Virginia's continental units. It was formed primarily of rifle companies (as opposed to the normal arrangement of seven musket companies and three rifle companies that all of the other Virginia regiments consisted of).

---

[6] Charles H. Lesser, ed., "A General Return of the Continental Forces, May 21, 1777," *The Sinews of Independence, Monthly Strength Reports of the Continental Army,* (University of Chicago Press, 1976), 46

[7] Ibid.

The 11th Virginia was commanded by Colonel Daniel Morgan of Frederick County, an officer who had demonstrated his support for the American cause almost two years earlier by leading one of Virginia's two original rifle companies northward to Boston in the summer of 1775. Morgan revealed his natural leadership and courage in Benedict Arnold's epic march to Quebec through the Maine wilderness and subsequent attack upon that fortress city in late 1775.

During the doomed American assault on Quebec, Captain Morgan led the Americans forward upon the loss of Colonel Arnold (who was shot in the leg) and was one of the last to surrender to the British. Held as a prisoner of war for eight months, Morgan returned to Virginia in the fall of 1776 and learned that General Washington himself had requested that a regiment of Virginians be saved for Morgan to command.

Colonel Morgan was officially exchanged in early 1777 and placed in charge of the 11th Virginia Regiment, a unit ideally suited for him due to its high number of rifle companies. Unfortunately, five of Morgan's companies that were already in the field as independent units were captured when Fort Washington fell to the British in November 1776, so Colonel Morgan scrambled over the winter of 1776-77 to replace these lost companies. The result was a regiment consisting of over 300 effective officers and men in late May 1777, one company of which were actually Pennsylvania riflemen, but the rest Virginians, most with rifles but at least two companies with muskets.[8] Although many of Morgan's troops were inexperienced, some of his officers, including

---

[8] John Dorman, ed., "Eden Clevenger Pension Application," *Virginia Revolutionary Pension Applications*, Vol. 20, (Wash. D.C., 1958 -), 9 and Lesser, 46

William Heth, Christian Febiger, Peter Bruin and Charles Porterfield, were veterans of Arnold's expedition and would prove to be, like Colonel Morgan himself, exceptional commanders.

The 15th Virginia Regiment was the least experienced of General Woodford's four regiments and had still not arrived in camp by May of 1777. Its commander, Colonel David Mason, would never actually arrive, resigning his commission in the summer before he left Virginia. When the regiment did join General Woodford's brigade, it was commanded by Lieutenant Colonel James Innes of Williamsburg.

Although the American army grew significantly over the spring of 1777, General Washington remained anxious about his troop strength and pressed his brigade commanders with orders to, "*enquire minutely into the State & Condition of your Brigade....*"[9] A large part of overseeing the condition of his brigade involved disciplining the troops, for which General Woodford formed courts of inquiry and brigade courts martial at their encampment in Middlebrook to address infractions of army regulations. The formation of fatigue details to supply and clean up after the brigade and guard detachments to maintain military order and security also fell to General Woodford and his fellow brigadier-generals.[10]

---

[9] Chase, ed., "Circular Instruction to the Brigade Commanders, 26 May, 1777," The Papers of George Washington, Vol. 9, 532

[10] William Heth, "Orderly Book of Major William Heth of the Third (sic) Virginia Regiment, May 15-July 1, 1777," *Virginia Historical Society Collections,* New Series, 11, 1892
    The author contends that this source is mis-identified because Major Heth served in the 11th Virginia until July 1777, after which he was transferred to the 3rd Virginia.

General Washington urged frequent formations for roll call and drill and repeatedly called upon the officers who were away from camp on furlough to return as soon as they could. Thousands of enemy troops in New Brunswick, New Jersey seemed poised to strike in early June, either against Washington's army (which was less than ten miles away in Middlebrook) or against the American capital, Philadelphia.

Prior to any British movement, and perhaps in anticipation of it, General Washington formed a new unit from among his riflemen in the army to serve as a reconnaissance force of light infantry. Washington selected Colonel Morgan of Woodford's brigade to command this select corps of riflemen, 500 strong, and Morgan chose the best marksmen he could find from among the Virginia, Pennsylvania, and Maryland troops (the three states that provided riflemen to the army).

General Woodford's brigade supplied about a quarter of the riflemen in the corps.[11] By mid-June Morgan's Rifle Corps was operating as an independent unit, one destined for great accomplishments.

While the American army steadily grew in strength among the Watchtung Mountains of New Jersey, General Howe and the thousands of British and Hessian troops under his command remained just a few miles away in New Brunswick, baffling General Washington with his inactiveness. Howe

---

[11] Note: Captain Thomas Posey of the 7th Virginia and Captain Gabriel Long of the 11th Virginia commanded two of Morgan's eight rifle companies in the rifle corps. Their men were drawn exclusively from General Woodford's brigade. See: Michael Cecere, *They Are Indeed a Very Useful Corps: American Riflemen in the Revolutionary War*, (Heritage Books, 2006), Appendix, "An Incomplete List of Riflemen who Served in Colonel Daniel Morgan's Rifle Corps in 1777," Payrolls transcribed by Joseph Craig, Saratoga NPS, 199-202

finally moved his troops westward, towards Washington on June 13th. The British commander had no interest in attacking the fortified American positions in the mountains, however, and sought instead to draw Washington's army into the open. The American commander refused to accommodate General Howe and sent only light parties, including Morgan's rifle corps, forward to skirmish with the British. After nearly a week of inconclusive fighting, Howe disengaged and withdrew back to New Brunswick. Then, inexplicably, Howe abandoned New Brunswick and marched to Perth Amboy, where it appeared he intended to cross over to Staten Island.

Just as General Howe had hoped, the surprising British withdrawal emboldened General Washington to send Colonel Morgan's rifle corps and General William Alexander (Lord Stirling's) brigade of Pennsylvanians forward to harass Howe's retreat. On June 25th the British commander seized the opportunity to strike the pursuing troops, who were now in the open. Marching through the night around Stirling's left flank to encircle the Americans and cut them off from the mountains, Howe's movement was detected by the Americans before he could close the trap. Heavy fighting and speedy maneuvering by the Americans enabled them to escape from Howe's trap and return to the relative safety of the Watchtung Mountains. Disappointed, General Howe abandoned further operations in New Jersey. He had a new plan in mind.

General Woodford and his brigade did not participate in the fighting that occurred in late June. They remained with General Washington and the main American army several miles in the rear. Woodford devoted some of his time in camp and on the march to a regular correspondence with his friend, Edmund Pendleton. In one letter, Woodford apparently referred to himself as a fallen angel within the army. This was likely a reference to his resignation from the army the previous fall. Pendleton replied that

> You was a little wrong in your metaphor of a Fallen Angel. They fall never to hope again, yours hath ascended the height he fell from and with equal rapidity.[12]

As much as General Woodford appreciated Pendleton's words of encouragement, it was the civilities Pendleton showed to Woodford's children that really touched the general. Pendleton commented on what he viewed as Woodford's excessive gratitude in one of his weekly letters to his friend.

> You and Mrs. Woodford rate too highly our common civilities to your boys. Were their Father unknown to us, his being abroad in the service of his Country would be a sufficient inducement to direct my regard to them, but when that Father is my most esteemed friend and I wish to cultivate with them a Family Friendship as in the Third Generation (for I had the pleasure of beginning it with their Grandfathers) I would if there was Occasion to much to serve them. They were well yesterday and the Family much devoted to you.[13]

---

[12] Mays, ed., "Edmund Pendleton to William Woodford, July 19, 1777," *The Letters and Papers of Edmund Pendleton*, Vol. 1, 217-218
[13] Mays, ed., "Edmund Pendleton to William Woodford, August 15, 1777," *The Letters and Papers of Edmund Pendleton*, Vol. 1, 219-220

Edmund Pendleton's weekly letters were undoubtedly a great comfort to General Woodford and he strove to reply to each letter, but this proved difficult once the American army broke camp in July.

General Howe's limited movements in 1777 had thus far perplexed General Washington, but by early July the American commander had concluded (hesitantly) that Howe intended to cooperate with a second British army (over 7,000 men strong) moving through New York from Canada under General John Burgoyne. Still unsure where Howe would strike, but aware that the bulk of his army was being loaded aboard transport ships, Washington shifted his army back to Morristown, New Jersey. He wanted to be in a better position to march north into New York to assist the American northern army upon confirmation of Howe's movement to join Burgoyne. Such confirmation never arrived, however. What ensued instead were weeks of contradictory reports of British ship sightings along the coast of New Jersey, Delaware, and Maryland that thoroughly baffled General Washington, keeping the American commander-in-chief and his army on constant alert.

Tour of the Jerseys

General Washington's assumptions regarding General Howe's intensions were mistaken, but it took weeks for the American commander to discover this because bad weather had forced the British fleet carrying General Howe's army southward far out to sea and out of sight of the coast. General Howe, with approximately 15,000 British and German troops, had sailed from New York in July and had disappeared over the horizon. General Washington believed this maneuver was a ruse and that they would soon reappear to sail up the Hudson River to cooperate with General Burgoyne marching through New York from Canada, so Washington ordered the American army to march northward, towards New York in mid-July. As General Washington was not certain that the New York Highlands was General Howe's true destination, his march north was done with little urgency. During the twelve days the army moved north, seven were spent in camp, usually waiting for the weather to improve. The five days of extended marches (scattered over the twelve days) averaged approximately 15 miles each day and took General Woodford's brigade from Morristown, New Jersey to Chester, New York.[14] Other American brigades were positioned closer to the border of New Jersey and New York.

On July 24th, General Washington received surprising intelligence that General Howe was actually sailing southward, most likely to Philadelphia via the Delaware River or Chesapeake Bay. Alarmed that his army was well out of position to challenge Howe, Washington ordered his several divisions to march to Philadelphia by the most expeditious route they could.[15] What transpired was a difficult forced march across New Jersey for the American troops.

Averaging more than twenty miles a day in the hot July sun, Captain John Chilton of the 3rd Virginia Regiment described the ordeal General Woodford's troops experienced.

---

[14] Michael Cecere, "The Diary of Captain John Chilton, July 12-24, 1775," *They Behaved Like Soldiers, Captain John Chilton and the Third Virginia Regiment, 1775-1778*, (Heritage Books, 2004), 114-116

[15] Frank E. Gizzard, Jr., ed. "General Washington to General Adam Stephen, July 24, 1777," *The Papers of George Washington*, Vol. 10, (Charlottesville, VA: University of Virginia Press, 2000), 399

> *As our March was a forced one & the Season extremely warm the victuals became putrid by sweat & heat – the Men badly off for Shoes, many being entirely barefoot and in our Regt. a [two] minute inspection was made into things relative to necessaries that the Men could not do without, which they were obliged to throw away.*[16]

In just over a week General Woodford and his brigade marched across New Jersey and crossed the Delaware River, camping on the outskirts of Philadelphia. Once again, however, General Washington found himself uncertain how to proceed. Reports that General Howe's fleet had disappeared over the horizon again made it impossible for Washington to determine the British army's destination. General Washington feared that General Howe had doubled back and was actually heading to the Hudson River, so he prepared to move the American army back into New Jersey.[17] They had only marched a short distance on August 10th, however, when General Washington received word from John Hancock, the President of Congress, that a large fleet had been briefly spotted off the coast of Maryland before it disappeared over the horizon again. If the report was true and General Howe was sailing south, he might be headed for the Chesapeake Bay to land his army below Philadelphia to strike the American capital from the south. Or perhaps Howe was heading to South Carolina. If so, there was little General Washington

---

[16] Cecere, "The Diary of Captain John Chilton, July 27, 1777, *They Behaved Like Soldiers...*, 117

[17] Grizzard Jr., ed., "General Washington to Samuel Washington, August 10, 1777," *The Papers of George Washington*, Vol. 10, 581

could do to reach the Carolinas in time to help. Unsure what to do, Washington remained a few miles northeast of Philadelphia and waited for further word.

Ten days passed without further intelligence on the British fleet and General Washington grew impatient. At a council of war attended by General Woodford and the other division and brigade commanders, it was agreed that since no new intelligence had arrived as to his whereabouts, General Howe was likely sailing to Charlestown, South Carolina. The council agreed that since there was little the army could do in time to assist Charlestown, they should march to New York to reinforce the American northern army against General Burgoyne's force of 7,000 British and Hessian troops.

Upon this recommendation, General Washington ordered the army to prepare to march, but within hours of these orders, news arrived that the British fleet had finally been sighted passing the Virginia Capes. It now appeared that Chesapeake Bay was the destination of General Howe, and if so, that meant that Philadelphia was likely his target.

General Washington immediately adjusted his orders, the army was to march to Philadelphia. While Washington's troops marched through the city and into Delaware, General Howe's troops sailed up the Chesapeake Bay and landed at Head of Elk near the Maryland and Delaware border. General Washington posted the bulk of his troops just south of Wilmington, Delaware and sent a 1,100 man detachment of light infantry (drafted from the several brigades including General Woodford's) further south under General William Maxwell to challenge any British advance.

In early September a sharp engagement at Cooches Bridge, Delaware occurred between General Washington's light corps and General Howe's advance guard resulting in the American light troops withdrawing towards Wilmington. General Howe did not pursue the retreating Americans, however, but marched his army northward past Washington's right flank. Howe encamped at Kennett Square in Pennsylvania while Washington re-positioned his troops at Chads's Ford, several miles to the east of Howe, in order to block the British from advancing on Philadelphia. A modest creek, the Brandywine, separated the two armies, each approximately 15,000 strong.

Battle of Brandywine

General Washington centered the deployment of his army at Chads's Ford, posting General Anthony Wayne's 2,000 Pennsylvanians to defend the crossing. General Nathanael Greene's 2,500 Virginians (under General George Weedon and General Peter Muhlenberg) defended the ground to Wayne's left, immediately south of the ford. They were supported by General Francis Nash's Carolina Brigade (1,500 strong). General Adam Stephen's division of Woodford's and Scott's brigades, 1,500 strong, as well as General William Alexander (Lord Stirling's) division of Pennsylvania and New Jersey troops, also 1,500 strong, were posted in reserve on a hill overlooking Brandywine Creek, just north of Chads's Ford. About a mile north of their position was General John Sullivan with a division of 1,100 Maryland troops.[18] They guarded another ford as well as policed the American right flank against any surprise by General Howe. With battle imminent, General Washington posted General Maxwell's

---

[18] Thomas McGuire, *The Philadelphia Campaign: Brandywine and the Fall of Philadelphia,* Vol. 1, (Stackpole Books, 2006), 170-171, 197

1,000 man light corps across Brandywine Creek to watch the enemy's movements and harass them if and when they moved against the Americans.

Maxwell's men did not have long to wait to detect movement of the enemy. General Howe ordered a column of his army forward at daybreak on September 11th, and shortly after dawn gunfire erupted in the countryside west of Brandywine Creek.

While Maxwell's light corps engaged the advance troops of General Howe's army, the bulk of the British forces, led by General Howe himself, marched along a circuitous route to gain Washington's right flank and replicate Howe's rout of Washington at Long Island in 1776. General Washington anticipated Howe's movement and sought to exploit the division of Howe's army by ordering an attack upon what he assumed was a much smaller enemy force to his front. Just as his troops moved forward to cross the creek, however, conflicting intelligence reports on Howe's movements created uncertainty in Washington about the actual size of the enemy to his front, so he halted the attack until he could better determine the situation his army faced. By the time General Washington ascertained that Howe had indeed divided his army and was moving against his right flank, the opportunity to exploit Howe's move had vanished and Washington scrambled to protect his right flank.

With more than half of the British army bearing down on his right and rear, Washington ordered General Stephen and General Stirling to rapidly march their divisions to the heights near Birmingham Meeting House in order to head off the British. General Sullivan was ordered to follow with his division and assume overall command of the redeployed right wing of Washington's army.

The march route from Brandywine Creek to the heights near Birmingham Meeting House, with its narrow, winding roads over steep, wooded hills and deep ravines tested the stamina of Washington's troops. They doggedly pushed on for three miles to the village of Dilworth, then swung north towards a hill overlooking Birmingham Meeting House and a road running north to Osborne Hill, where General Howe and the British army had paused to rest after their long march around the American army. General Stephen deployed his troops upon a large, cleared, hill just west of the Birmingham Road. General Scott's brigade formed on the left of Woodford's brigade, and General Stirling's division formed to Scott's left.

General Woodford's brigade thus held the right flank of the American line, which meant he had no support on his right side. To alleviate his concern for his exposed right flank, General Woodford ordered his most experienced regiment, Colonel Thomas Marshall's 3rd Virginia, only 170 strong, to post themselves in an orchard and among the outbuildings of a farm just north of Birmingham Meeting House, several hundred yards in advance and to the right of the American line. From that position they could detect, and possibly stop, any effort of the enemy to get around Woodford's right flank.

The arrival of General Sullivan's division on the left of the American line, however, caused General Stephen and General Stirling to shift their divisions several hundred yards to the right. General George Weedon, who did not witness the engagement at Birmingham Heights but likely discussed it with officers who participated in it, explained the impact this shift to the east had on General Woodford's deployment.

> *In making this Alteration, unfavorable Ground, made it necessary for Woodford to move his Brigade 200 Paces back of the Line & threw Marshall's Wood in his front.*[19]

The troops of the 3rd Virginia were no longer on the right of General Woodford's line, they were to his front, exposed but protected by an orchard and stone wall. A mile to their front were approximately 8,000 British and German troops, eager to attack after their day long march, while several hundred yards behind the 3rd Virginians waited about half that number of American troops, determined to offer spirited resistance.[20]

General Howe ordered his advance troops of German jaegers (riflemen) and British light infantry forward from Osborne's Hill shortly after 3:00 p.m. Howe's main body followed about 45 minutes later, marching down Osborne's Hill in several columns before deploying into battle lines as they drew closer to the American line.

In the orchard and farm adjacent to the Birmingham Meeting House, Colonel Marshall's 3rd Virginians waited anxiously for the enemy to come into effective range (approximately 100 yards for muskets, 200 yards for rifles). Captain Johann Ewald, in command of a company of jaegers, led his men straight at the Virginians and described the initial contact.

---

[19] Bob McDonald, transcribed, "Brigadier General George Weedon's Correspondence Account of the Battle of Brandywine, September 11, 1777," Manuscript is held by the Chicago Historical Society

[20] McGuire, 199

> *About half past three I caught sight of some infantry and horsemen behind a village on a hill in the distance. I drew up at once and deployed...I reached the first houses of the village with the flankers of the jagers, and Lt. Hagen followed me with the horsemen. But unfortunately for us, the time this took favored the enemy and I received extremely heavy small-arms fire from the gardens and houses, through which, however, only two jagers were wounded. Everyone ran back and I formed them behind the fences and walls at a distance of two hundred paces from the village.*[21]

The intensity of the Virginians resistance surprised the British advance troops and they withdrew for the cover of a fence and embankment along the road where they waited for reinforcements. Colonel Marshall's Virginians held firm and waited as well.

Captain Ewald was eventually joined by hundreds of British light infantry (as well as the rest of Howe's main body to his right) whose arrival triggered a withdrawal of the 3rd Virginians to a stone wall at the Birmingham Meeting House. Shielded by this wall, the battle tested Virginians maintained such a heavy fire on the enemy that the British troops advancing upon them balked at a frontal assault and maneuvered around their flanks instead.

Praise for the conduct and bravery of the 3rd Virginians at the orchard and Meeting House was universal. General Weedon noted that Colonel Marshall and his men,

---

[21] Johann Ewald, *Diary of the American War: A Hessian Journal*, (New Haven & London: Yale University Press, 1979), 84-85 Translated & edited by Joseph P. Tustin

> *Received the Enemy with a Firmness which will do Honor to him & his little Corps, as long as the 11th of Sepr. is remembered. He continued there ¾ of one Hour, & must have done amazing execution.*[22]

Henry "Light Horse Harry" Lee wrote years later in his memoirs of the war that the 3rd Virginia,

> *Bravely sustained itself against superior numbers, never yielding one inch of ground and expending thirty rounds a man, in forty-five minutes.*[23]

Such resistance though, came at a heavy cost. Over a third of the regiment was lost in the fight, killed, wounded or captured, and the survivors barely escaped.[24] With enemy troops closing in on both sides of the Meeting House wall, the remnants of the 3rd Virginia, led by Colonel Marshall on foot (his horse having been shot out from under him) scurried southward to re-join General Woodford's brigade and continue the fight.

Woodford's troops, with General Scott's Virginians posted to their immediate left, were deployed in a strong position a bit south of the rest of the American line (Stirling's and Sullivan's divisions). A British light infantry officer observed that, *"The position the enemy had taken was very strong indeed -- very commanding ground, a wood on their rear and*

---

[22] McDonald, transcribed, "Brigadier General George Weedon's Correspondence Account of the Battle of Brandywine, September 11, 1777"

[23] Henry Lee, *The Revolutionary War Memoirs of General Henry Lee*, (New York: Da Capo Press, 1979), 89-90
   Originally published in 1812

[24] McGuire, 215

flanks, a ravine and strong paling [fence] *in front."*[25] This British officer was likely describing the position of General Woodford's troops on the far right of the entire American line. When General Stephen shifted his division right to accommodate the arrival of General Sullivan's division, low ground forced Stephen to deploy his troops 200 yards further south, along higher ground and woods. This turned out to be fortunate for the Americans as the difficult terrain to their front slowed the British advance and helped Woodford and Scott hold firm. Captain John Montresor, General Howe's chief engineer, offered his own description of the terrain defended by Woodford and Scott's troops.

> *The ground on the left being the most difficult the rebels disputed it with the Light Infantry with great spirit, particularly their officers, this spot was a ploughed hill; and they, covered by its summit and flanked by a wood; however unfavorable the circumstances* [the ardour of the British lights] *was such that they pushed in upon* [the rebels] *under a very heavy fire.*[26]

It would take more than a direct frontal assault to pry Stephen's division from its strong position. Unfortunately for the Americans, such was not the case on the opposite side of their line, where General Sullivan's division was unable to get into position before the British attacked. While some of Sullivan's men fought well, many fell into disorder at the arrival of the British and withdrew to the rear. General Sullivan, who had taken position near a battery of five cannon

---

[25] McGuire, 216
[26] Ibid.

in the center of the entire American line to superintend the fight, sent his aides to reform his troops, but they had little success and most of Sullivan's troops withdrew in disarray.

With British troops to their front and now upon their exposed left flank, the pressure on General Stirling's men in the center of the American line increased and they also began to give way. To their right though, remained Stephen's division with General Woodford and Scott at the head of their brigades.

The fight between Woodford and Scott's Virginians and the British light infantry troops to their front was severe, made more so upon the British by two of Woodford's cannon that played very effectively upon the British. A British officer who experienced the cannon fire recalled that,

> *There was a most infernal Fire of Cannon & musketry – smoak – incessant shouting – incline to the right! Incline to the Left! – halt! – charge!...The trees* [were] *cracking over ones head. The branches riven by the artillery, the leaves falling as in autumn by the grapeshot.*[27]

Woodford's Virginians were well served by their cannon which, along with heavy musket and rifle fire, initially kept the British and German troops to their front at bay. A German officer noted that, "*The small arms fire was terrible, the counter-fire from the enemy, especially against us, was the most concentrated.*"[28]

---

[27] Ibid.
[28] McGuire, 237

The collapse of the American left flank and center, however, left Stephen's Virginians in an impossible situation. A British officer described the final assault on Woodford's position.

> *They stood the charge till we came to the last* [fence]. *Their line then began to break, and a general retreat took place soon after, except for their guns, many of which were defended to the last, indeed several officers were cut down at the guns.*[29]

Although American reinforcements under General Weedon and Muhlenberg arrived in time to screen the American withdrawal, the fighting for General Woodford and his men was over for the day and General Washington's army withdrew towards Philadelphia to regroup. General Woodford's brigade, particularly the 3rd Virginia Regiment had fought heroically and suffered heavy casualties for their effort. General Woodford himself was included among the casualties, suffering a gunshot wound to his left hand that threatened to leave him permanently disabled.[30]

---

[29] Ibid., 238
[30] Mayes, ed., "Edmund Pendleton to William Woodford, November 8, 1777," *The Letters and Papers of Edmund Pendleton*, Vol. 1, 234

Battle of Brandywine

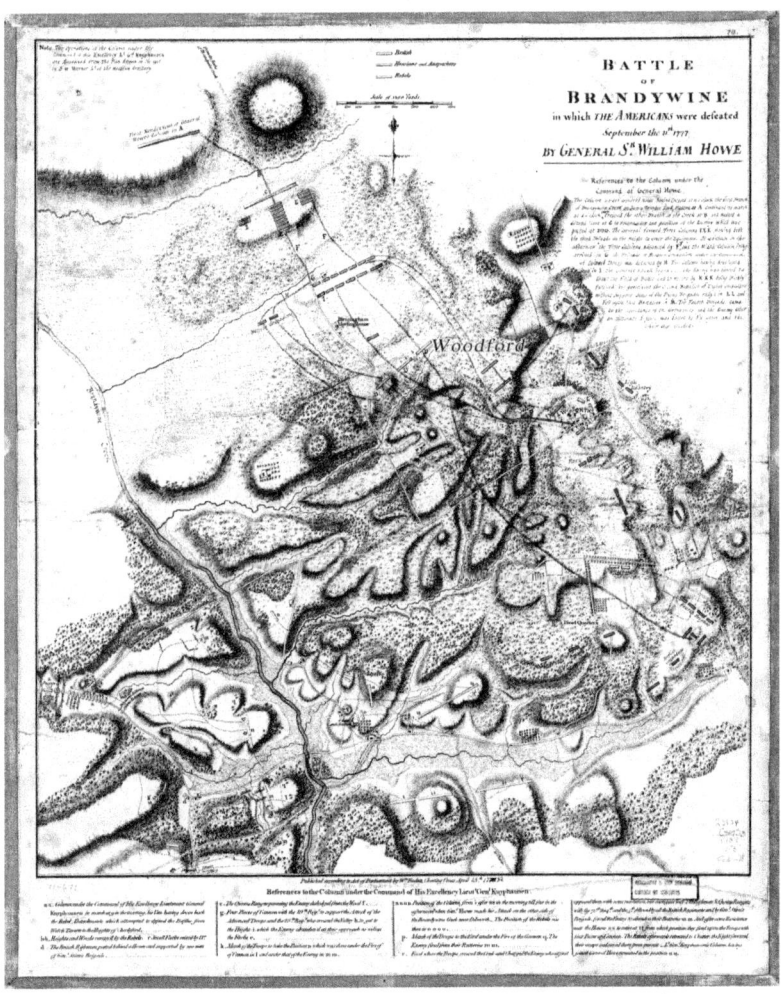

Chapter Eight

"When Rank is Once Given...the Party in Possession of it in Most Cases is Unwilling to Give It Up."

1777-1778

General Woodford temporarily left his brigade to recover in the Moravian settlement of Bethlehem, Pennsylvania where a supply depot and hospital had been established. He was joined by a young French volunteer who had himself been wounded at Brandywine, shot through the calf. The Marquis de LaFayette had arrived in America against the wishes of his superiors in France, eager to join the fray and gain glory. Although General Washington was initially resistant to offer the young French nobleman a position in the American army, LaFayette found himself amidst the fighting at Birmingham Heights as an observer. General Woodford and LaFayette spent four weeks in Bethlehem recovering from their wounds, during which each developed a favorable impression of and friendship with the other.

As Bethlehem was also the site in which a large amount of American supplies were stored, there were a number of troops in town as guards, not to mention those troops who had recovered from their injuries but who lingered in town. General Washington requested General Woodford, as the ranking officer in Bethlehem, to oversee the conduct of these troops and send as many as he could back to the main army.[1]

Woodford updated Washington on his efforts in late September, reporting that there were less than 200 men fit for duty at Bethlehem of which Woodford sent 25 of the best back to rejoin Washington's army. General Woodford lamented to the commander in chief that

> *Many complaints are Daily made by the Country people of Robberys and other disorders committed by these Scum of the Army. I have taken every method to detect the delinquents & bring them to Justice, but they have hitherto proved affectual.*[2]

He promised to make an example of the first soldier caught participating in such conduct and encouraged the country folk to fire upon any soldier snooping about their houses or farms.[3] On a lighter note, Woodford informed Washington that both he and the Marquis de LaFayette were recovering nicely and that the young Frenchman was in high spirits.

LaFayette was apparently impressed by General Woodford, for in a letter to Washington written in mid-October, the Marquis boldly requested command of a division of Virginians, *"with general Woodfort would be the most agreeable* [situation for me]."[4] LaFayette had learned that

---

[1] Philander D. Chase and Edward G. Lengel, eds., "General Washington to General Woodford, September 26, 1777," *The Papers of George Washington*, Vol. 11, (Charlottesville, VA: University of Virginia Press, 2002), 329

[2] Chase and Lengel, eds., "General Woodford to General Washington, October 2, 1777," *The Papers of George Washington*, Vol. 11, 371

[3] Ibid.

[4] Chase and Lengel, eds., "Marquis de LaFayette to General Washington, October 13, 1777," *The Papers of George Washington*, Vol. 11, 507

Woodford's division commander, General Adam Stephen, was being investigated for improper conduct in the Battle of Germantown (accused of being drunk during the attack that occurred outside of Philadelphia while Woodford and LaFayette were recuperating) and so the young Frenchman took the opportunity to lobby for Stephen's command. LaFayette's efforts would eventually pay off, for he did indeed assume command of Stephen's division upon the latter's court martial and subsequent cashiering from the army.

Woodford Rejoins the Army

When General Woodford and LaFayette re-joined the army just north of Philadelphia at Whitemarsh, Pennsylvania in late October, General Stephen's situation had yet to be determined. General Washington was far more concerned with how to respond to General Howe's capture of Philadelphia and the American army's inexplicable collapse at the Battle of Germantown in early October. Reports of General John Burgoyne's surrender at Saratoga, New York (with an entire British army) were very welcome in camp and lifted American spirits, but the fact remained that the American capital had fallen to the enemy and Washington's army had experienced nothing but defeat in the fall of 1777.

General Washington wished to reverse this, but a war council in late October, just before General Woodford rejoined the army, advised against an attack on the British in Philadelphia.[5] As a result, Washington's army of approximately 11,000 troops remained at Whitemarsh, waiting

---

[5] Frank E, Grizzard, Jr. and David R. Hoth, eds., "War Council, October 29, 1777," *The Papers of George Washington*, Vol. 12, (Charlottesville, VA: University of Virginia Press, 2002), 46

for reinforcements from the victorious Northern army in New York to arrive.[6]

While they waited, American troops in two forts on the Delaware River south of Philadelphia resisted British attempts to capture the forts and open river navigation all the way to Philadelphia. If the British succeeded in taking the forts, no further obstacles would stand in the way of the British navy reaching Philadelphia to supply General Howe's army. The American garrisons in the forts fought heroically, but could not prevail and their commanders evacuated them in mid-November in the face of certain defeat.

The situation for the main American army at Whitemarsh, less than a day's march north of Philadelphia, was far less critical. With little enemy activity north of Philadelphia, General Washington's army struggled more with the daily challenge of supplying and disciplining the troops than any serious threat by the enemy. Within days of his return to his brigade, General Woodford had a testy exchange with the army clothier general, James Mease, over the supply of clothing for his troops. Woodford complained to General Washington about the incident and the commander-in-chief intervened with a letter to Mease.

> *Genl Woodford complains that he lately wrote you a polite letter requesting necessaries for his Brigade, which he sent by an Officer, to which he says you only returned him a rough verbal Answer, without complying with his demand even in part. As General Woodford is an exceeding good Officer and one who I think would*

---

[6] Ibid.

not make extravagant or unnecessary demands, I could wish you would clear up the matter to his satisfaction.[7]

The arrival of reinforcements from the north in late November emboldened General Washington to once again consider an attack on Philadelphia. General John Cadwalader of Pennsylvania drafted a plan for such an attack and Washington circulated it among his generals for their input. Reaction was split, with some believing an attack was too risky while others, including General Woodford, believing that, *"something should be attempted by this Army before it retires into Winter Quarters, both for its own* [credit] *& the support of our paper currency."*[8] Woodford added that the army was not likely to get any stronger than it was at that moment, and the detached nature of the enemy, who had sent a large force into New Jersey to forage, *"promises a fairer prospect of Success, than is likely to present itself again whilst we are able to continue in the Field."*[9]

With his officers divided on whether to attack Philadelphia and his own instincts urging caution, General Washington resolved to remain at Whitemarsh in anticipation of a push against him by General Howe.[10] Washington's hunch proved right as the British marched out of Philadelphia towards Whitemarsh on December 4th with over 10,000 troops. The Americans braced for combat behind hastily erected fortifications in the low hills of Whitemarsh but a general

---

[7] Grizzard, Jr. and Hoth, eds., "General Washington to James Mease, November 12, 1777," *The Papers of George Washington*, Vol. 12, 228

[8] Grizzard, Jr. and Hoth, eds., "General Woodford to General Washington, November 25, 1777," *The Papers of George Washington*, Vol. 12, 404

[9] Ibid.

[10] Grizzard, Jr. and Hoth, eds., "General Washington to General Greene, November 26, 1777," *The Papers of George Washington*, Vol. 12, 407

engagement never developed. Colonel Daniel Morgan's Rifle Corps, recently returned from New York, successfully ambushed a large body of the enemy on the left flank of the line, but the bulk of the troops from both sides did not engage and after a weeklong standoff, General Howe withdrew back to Philadelphia, bringing the campaign season to a close.

It was time for General Washington to select a location for the army to spend the winter. Unlike the previous winter, in which Washington's troops were spread out among New Jersey's Watchtung Mountains, the American commander wanted to keep the army relatively intact. To do this he needed a location that could accommodate the army's needs. Whitemarsh would not work because it was too close to Philadelphia, just a few hours march from the British army. Washington needed someplace further away (to allow for a proper warning if the British did try to attack over the winter) but also defensible and logistically sustainable. He settled on Valley Forge, about 25 miles northwest of Philadelphia, and marched the army there on December 19th.

Valley Forge

Valley Forge provided strong terrain on which to resist a possible British attack. The Schuylkill River protected the American left flank and a steep hill, called Mount Joy, covered their rear. Although the front and right flank of the encampment possessed few natural barriers, the open terrain made an attack from those directions very hazardous.

General Washington ordered that log huts and earthworks be constructed immediately, but it took nearly a month before the entire army was adequately sheltered. Two lines of earthworks were built. The outer line extended along a ridge from the Schuylkill River to the foot of Mount Joy. Most of the army was stationed along this line in rows of huts behind the fortifications. An inner defense line was built along Mount Joy. It also ran to the river. General Woodford's brigade was posted at the southern base of Mount Joy to defend a gap between the outer and inner lines. Woodford established his quarters in a farmhouse (that still stands) about a mile and a half south of his brigade's position.[11]

A troop return for the army for January 1778 reported Woodford's troop strength as 1,287 among his four regiments, but of these, only 231 were present and fit for duty.[12] Over two thirds of his brigade was unfit for duty due to sickness, lack of clothing, or they were away on furlough. Another two hundred men were on detached service, many with Colonel Morgan's Rifle Corps. Such numbers were similar for almost every brigade at Valley Forge, the long campaign had been very hard of Washington's troops, and those still fit for duty struggled to construct huts to shelter themselves and their comrades from the winter.

The suffering of the men due to lack of proper clothing was compounded with a dire lack of provisions at the beginning of the encampment and periodically throughout the winter and spring. On December 22nd, General James Varnum of Rhode Island reported to General Washington:

> *Three Days successively, we have been destitute of Bread. Two Days we have been intirely without Meat. –It is not to be had from Commissaries. –Whenever we procure Beef, it is of such a vile Quality, as to render it a poor Succedanium for Food. The Men must be supplied, or they cannot be commanded.* [13]

---

[11] The Tredyffrin Easttown Historical Society, places Woodford's quarters about a mile and a half south of his brigade's encampment.

[12] Lesser, ed., "Return of the Continental Army, January 3, 1778," *The Sinews of Independence: Monthly Strength Reports of the Continental Army*, 58

[13] Joseph Lee Boyle, "General Varnum to General Washington,

General Washington concurred with General Varnum and warned Congress that the very existence of the army was at stake if the supply situation did not improve immediately.

> *I do not know from what cause this alarming deficiency, or rather total failure of Supplies arises: But unless more vigorous exertions and better regulations take place in that line and immediately, This Army must dissolve.*[14]

Rank Controversy

If supply problems were not enough for General Washington to deal with, Congress sparked a new crisis among the officer corps of the army when they attempted to settle the seemingly endless dispute over rank and order of precedence for promotions among many of the officers in the army. Scores of American officers were dissatisfied with their rank and placement in the officer hierarchy and General Woodford was one of these officers. He had resigned from the army in 1776 because two officers of lower rank at the time (Colonel Mercer and Colonel Stephen) had been promoted to brigadier-general ahead of him, but in early 1777 Woodford agreed to return to the army when Congress offered him a brigadier-general's commission. Unsatisfied with his placement (third) among the four new Virginia brigadier generals appointed in 1777 (Weedon, Muhlenburg, Woodford, and Scott) all of who had ranked beneath Woodford prior to

---

22 December, 1777," *Writings from the Valley Forge Encampment of the Continental Army,* Vol. 1, (Bowie: Heritage Books Inc., 2000), 2

[14] Grizzard Jr., and Hoth, eds., "General Washington to Henry Laurens, 22 December, 1777," *The Papers of George Washington,* Vol. 12, 667

his resignation in 1776, Woodford and several of his allies in Congress let it be known for much of 1777 that he desired Congress to restore him to his proper place (first) among the Virginia brigadier generals.

In addressing a similar dispute over rank among Pennsylvania officers in August of 1777, Congress approved a Board of General Officers proposal (ironically of which General Woodford and General Weedon were members) to establish the rank and precedence of the Pennsylvania officers according to, "*that standing they held in the army immediately before their present commissions.*"[15] Several months later, in November, Congress applied this same measure to other officers, including General Woodford, who sought adjustments to their rank. Henry Laurens, the president of Congress, instructed General Washington on November 29th, to regulate the ranks of Generals Benedict Arnold, Charles Scott, and William Woodford based on the arrangement used for the Pennsylvania officers.

General Washington believed there would be little opposition to an adjustment of General Arnold and General Scott's ranks, as most recognized that their situations deserved to be corrected, but General Woodford's situation was different, he had voluntarily left the army (against Washington's advice) so there was little sympathy among Woodford's fellow officers for his argument that he deserved a higher placement among the brigadier generals because he served as a colonel in the army well before the other officers had. General Washington delayed addressing General Woodford's situation until the army marched to Valley Forge.

---

[15] Ford, ed., "November 12, 1777,"*Journal of the Continental Congress*, Vol. 9, (Washington, D.C.: Government Printing Office, 1907), 896

When he did address it on December 22, Washington sought further guidance from Congress as to how to resolve General Woodford's situation. He also warned Congress of the importance of settling the question of rank once and for all and as quickly as possible.

> *At the time General Woodford was appointed he held no rank in the Army. His claim is to rank before such of the Gentlemen* [Muhlenberg, Scott, Weedon] *appointed when he was, as* [they] *were younger Colonels than himself. If this was intended by Congress...and I presume it was, or it can have no Operation as to him – An Explanatory and directory Resolve would answer every purpose, as from that a Commission might be filled up agreeable to their views respecting him. I wish this business to be determined on as early as possible...as no subject can be more disagreeable or injurious to the service than that of contested rank.*[16]

General Washington, who was undoubtedly annoyed at the mess Congress had dumped in his lap, realized the uproar a change in Woodford's rank would create among his fellow officers and lectured Congress to be more careful in the future when addressing the issue of rank in the army.

> *I trust Congress will be more guarded in future. They may not be so intimately acquainted with them as I am; But they may be assured, there are none of a more fatal and injurious tendency, when rank is once*

---

[16] Grizzard Jr., and Hoth, eds., "General Washington to Henry Laurens, 22 December, 1777," *The Papers of George Washington,* Vol. 12, 668

> given, no matter upon what principle, whether from mistake or other causes, the party in possession of it in most cases is unwilling to give it up.[17]

This was certainly the case regarding General Woodford's attempt to rise above Generals Muhlenberg and Weedon. Just as General Washington expected, the other brigadier generals from Virginia strongly objected to any arrangement that placed General Woodford above them. It was nothing personal, they claimed, they held General Woodford in the highest esteem and considered him an intimate acquaintance. It was the principle of honor, they argued that forced them to oppose the loss of their status through no fault of their own.[18]

Realizing the turmoil they had caused and the dilemma they were in, Congress formed a committee to go to Valley Forge to settle the dispute, but they did not address the issue until mid-February and after several deliberations, handed the problem off to yet another board of general officers. By this time, General Weedon had returned to Virginia on furlough and General Woodford had offered his resignation in frustration to expedite the matter. In doing so, Woodford argued his case one last time.

> *Dear Genl*
> *It is with the greatest reluctance that I think myself obliged to add to the many things that I know perplex your Excellency at this time. But the claim I made of being reinstated in my Rank in the line of my own State,*

---

[17] Ibid.
[18] Harry M. Ward, *Duty, Honor or Country: General George Weedon and the American Revolution,* (Philadelphia, PA: American Philosophical Society, 1979), 124-125

appears to be as far from being settled as ever. I have waited with patience these three Months, & done duty in the Line under Genls Muhlenberg & Weedon, contrary to my resolution when I enter'd the Service, the latter, who appears to be the only person that objects to this piece of Justice being done me, is gone to Virginia, from whence in all probability, he will not return till the active part of the campaign commences – I think I foresee this affair must again go back to Congress before it can be finally settled.... I...have remain'd satisfy'd to waite the determination of this Committee, who now inform me it is refer'd to a Board of Genl Officers.

I conceive there would be no impropriety in that Boddy doing me Justice...a Board of Genl Officers may perhaps consider my condescending to Serve the last Campaign under these Gentlemen as an acquiescence on my part, & a bar to my present claim – but as I have observed to the Commitee, it never can be too late to do Justice to an individual when a Publick Boddy are convinced he has been injured.

I have already inform'd your Excellency that nothing will induce me to remain longer in the Continental Line under the command of those Gentlemen, who were called into service long after me, & in Inferior Ranks. I must therefore ask your permission to retire, which I assure you I do with great reluctance.[19]

---

[19] Edward G. Lengel, ed., "General Woodford to General Washington, February 19, 1778," *The Papers of George Washington*, Vol. 13, (Charlottesville, VA: University of Virginia Press, 2003), 601-602

General Washington replied to Woodford with deep regret at his request to resign and with an assurance that the dispute would not be sent back to Congress, but rather, the decision of the board of general officers would be final. Washington added that the board could be convened immediately, to which General Woodford heartily approved and withdrew his request to resign.[20]

The board of general officers met in early March to consider Woodford's case but refused to pass final judgement on the matter. They did, however, recommend to the Congressional Committee still at Valley Forge that Congress decide in General Woodford's favor. The committee referred this recommendation to Congress, who finally resolved on March 19th, that

> *General Washington call in and cancel the commissions of Brigadiers Woodford, Muhlenberg, Scott, and Weedon; and that new commissions be granted them; and that they rank in future agreeable to the following arrangement, Woodford, Muhlenberg, Scott, Weedon.*[21]

General Woodford was thoroughly pleased by the decision, but General Weedon was not and resigned from the army in protest.

---

[20] Lengel, ed., "General Washington to General Woodford, February 19, 1778," and "General Woodford to General Washington, February 19, 1778," *The Papers of George Washington*, Vol. 13, 636-637

[21] Ford, ed., "Congressional Resolution, March 19, 1778," *Journal of the Continental Congress*, Vol 10, (Washington, DC,: Government Printing Office, 1908), 269

Furlough Home

It had been a difficult winter for the American army at Valley Forge and General Woodford's brigade, like many others, had dissolved to barely 100 men present in camp and fit for duty.[22] The advent of spring, however, promised better conditions and more troops and General Woodford took advantage of the improving situation to request a furlough home to Virginia, which was granted.

General Woodford left Valley Forge in late March carrying letters from General Washington to several people, including General Weedon, whose residence Woodford would pass in Fredericksburg. Weedon thus likely learned of Congress's decision to rank him below General Woodford from a letter delivered to him by General Woodford.

While the reception Woodford received from Weedon was in all likelihood a bit strained, he was received joyously at Windsor when he arrived home in early April, finding his wife and sons in good health. A year had passed since General Woodford had left Virginia and there was much to do to get his personal affairs in order. The restoration of his health was also a goal after the long winter at Valley Forge.

General Woodford was also tasked by General Washington with hastening the return to camp of both officers and men in Virginia that had returned earlier in the winter on furlough or who had recently been recruited or drafted. Woodford placed announcements in the Virginia gazettes ordering, on behalf of General Washington, that all continental officers and soldiers still in Virginia proceed to Valley Forge as soon as possible.[23]

---

[22] Lesser, 59
[23] David R. Hoth, ed., "General Washington to General Woodford, March 28, 1778," *The Papers of George Washington*, Vol. 14,

We know little about the two months General Woodford spent at home at Windsor. A letter from his friend Edmund Pendleton in early June hints that Woodford's health had improved considerably during his time home, so it was likely a restful stay.[24] He started back to Valley Forge around the end of May.

At some point during his return to the army, Woodford's father-in-law, John Thornton, died. He left no will, and General Woodford was tapped to help administer the large estate that was left behind. Although he was likely already back at Valley Forge, Woodford's name was included in an announcement in the gazettes in mid-June (along with another son-in-law, John Taliaferro) calling for all those with debts or claims against Thornton's estate to come forward as soon as possible and present their claims.[25] There was little that Woodford could actually do from Pennsylvania to help settle Thornton's accounts, however, as he was busy commanding his brigade in Pennsylvania.

Woodford did dissolve his partnership in a brewing enterprise in Fredericksburg likely because of the death of his father-in-law. Several buildings and lots in Fredericksburg, including a brew house and malt house as well as an adjacent tannery, two nearby lots, and three slaves engaged in the business of brewing, were all offered for sale.[26] General Woodford likely found managing these affairs from afar a significant challenge, made all the more so by the start of the active campaign season in June.

(Charlottesville, VA: University of Virginia Press, 2004), 346-347
[24] Mayes, ed., "Edmund Pendleton to William Woodford, June 6, 1778." *The Letters and Papers of Edmund Pendleton*, Vol. 1, 257
[25] Purdie's *Virginia Gazette*, June 19, 1778, 1
[26] Dixon and Hunter's *Virginia Gazette*, July 10, 1778, 4

Chapter Nine

"I am against risquing a Genl Attack."

1778

It is difficult to establish with certainty when General Woodford reached camp at Valley Forge on his return from furlough. Edmund Pendleton in a letter dated June 6, 1778, wrote that he expected Woodford had already reached headquarters, but we find no definitive evidence of Woodford's presence in camp until he attended a war council with his fellow officers on June 17th.[1] A lot had transpired with the army during Woodford's furlough. A steady stream of new soldiers, many drafted, as well as troops returning from furlough dramatically increased the size of the army. General Washington estimated that there were 12,500 troops fit for duty at Valley Forge in mid-June (although he acknowledged that the number capable of marching was closer to 11,000).[2] General Woodford's brigade had more than quadrupled in size since his departure in March and now counted over 650 officers and men fit for duty.[3]

---

[1] Mays, ed., "Edmund Pendleton to William Woodford, June 6, 1778," *The Letters and Papers of Edmund Pendleton*, Vol. 1, 257
and
Edward G. Lengel, ed., "War Council, June 17, 1778," *The Papers of George Washington*, Vol. 15, (Charlottesville, VA: University of Virginia Press, 2006),   415
[2] Lengel, ed., "War Council, June 17, 1778," *The Papers of George Washington*, Vol. 15,   415
[3] Lesser, "Monthly Return of the Continental Army...for May 1778," 68

Many of these men benefitted from the hard work of a recent addition to the army, Baron Freidrich von Steuben, an experienced officer from the Prussian army. With General Washington's blessing, Steuben had assumed the role of Inspector-General of the army and devised a new system of drill that was adopted by the entire army. Another important development that occurred during Woodford's absence was the arrival of news of an American alliance with France. The promise of French assistance boosted American morale and likely contributed to the decision of British General William Howe's successor, General Henry Clinton, to abandon Philadelphia and return to New York to regroup.

It was Clinton's preparation to leave Philadelphia that prompted General Washington to summon a war council on June 17th. Washington wished to know the opinion of his generals on how the American army should respond. He specifically sought advice on how aggressive the American army should be in response to what appeared to be the enemy's withdrawal from Philadelphia. Washington asked his generals to discuss the situation among themselves and then give him their individual opinions in writing.

General Woodford, like most of Washington's general, advised caution.

> *I am not for risqueing any enterprize against Philadelphia, & would recommend that the Army remain in its present position 'till your Excellency is well assured of the evacuation of the City, & the Rout the Enemy have taken. I think our present Strength will afford one Brigade more to be sent to the Jerseys, without exposeing us to any Danger, or at least that it might move to the*

Delaware, to be ready to cross, this I think would give confidence to the Militia of that State.... If such measures should be taken as well enable us to follow them, I am against risquing a Genl Attack, & to make a partial one must depend upon the Situation of the two armys, & circumstances that may happen hereafter.[4]

As it turned out, circumstances overtook the matter when General Washington learned the next morning that the British had indeed abandoned Philadelphia. He immediately issued orders for General Charles Lee (who had returned to the American army in a prisoner exchange after an absence of a year and a half) to lead six brigades from Valley Forge towards the Delaware River to pursue the enemy, estimated by Washington to be around 10,000 strong, into New Jersey.[5] General Woodford and the rest of Washington's army followed Lee early the next morning. They crossed the Delaware River two days later, where General Washington re-organized the army for possible battle. Washington divided the army into three parts, a right wing, a left wing, and a second line. General Woodford's brigade was attached to the right wing along with the brigades of Generals Scott, Poor, Varnum, Huntington, and the brigade of North Carolina

---

[4] Lengel, ed., "General Woodford to General Washington, June 18, 1778," *The Papers of George Washington*, Vol. 15, 469-470

[5] Lengel, ed., "War Council, June 17, 1778," *The Papers of George Washington*, Vol. 15, 415 and "General Washington to Henry Laurens, June 18, 1778," *The Papers of George Washington*, Vol. 15, 449

Note: General Charles Lee had been captured in late December 1776 by the British and had spent the last sixteen months on parole in New York City under British supervision. He was allowed to return to active duty upon his exchange for General Richard Prescott, a British general captured by the Americans in Rhode Island in the summer of 1777.

troops. General Lee was placed in command of this wing.[6] General Stirling commanded the left wing, composed of the troops from Pennsylvania and Massachusetts and General LaFayette commanded the second line with General Muhlenberg and Weedon's Virginia brigades as well as troops from Maryland and New Jersey.[7] Twenty five select men from each brigade were also ordered to join Colonel Daniel Morgan's rifle corps to bring his force to 600 men and allow him to better harass the enemy's flanks and rear.[8] While Morgan's troops raced ahead of the army to catch up with the British and, in conjunction with a brigade of New Jersey troops under General William Maxwell and militia under General Philemon Dickinson harass the rear and flanks of the retreating enemy column, Washington's main body followed as fast as they could, leaving their tents and heavy baggage to follow in the rear.

As the Americans drew closer to the British, General Washington called another war council to once again seek the opinion of his generals on whether a general engagement was advisable and if not, how they should proceed instead. None of Washington's officers favored an outright general attack, but they did support sending another detachment of light troops ahead of the main army to assist the other detachments already harassing the enemy.[9] The size of this detachment was a point of contention among Washington's officers. General Woodford and five other officers, including General

---

[6] Lengel, ed., "General Orders, June 22, 1778," *The Papers of George Washington*, Vol. 15, 493
[7] Ibid.
[8] Ibid.
[9] Lengel, ed., "War Council, June 24, 1778," *The Papers of George Washington*, Vol. 15, 520-521

Lee, favored a force of 1,500, but several others favored 2,500 while a few argued for 2,000.[10] General Washington was apparently unaware of the disagreement among the war council and ordered General Scott to lead 1,500 picked men forward.[11] When several officers informed Washington of their support for a stronger force, however, he sent another 1,000 men forward under General LaFayette and instructed the young Frenchman to take command of all of the advance detachments.[12]

With approximately 5,000 troops in several detachments now engaged as an advanced guard, General Lee, who had strongly argued against a general engagement, expressed his discomfort at not being in command of such a large part of the army. He was, after all, second in command of the army and thus, claimed Lee, entitled to lead such a force. General Washington appeased General Lee by sending him forward with yet another detachment (two brigades) with instructions to take overall command of the advance troops, but not to interfere with any enterprise that General LaFayette had already undertaken.[13] Washington hoped that this compromise would ease the egos of both officers.

---

[10] Lengel, ed., "General LaFayette to General Washington, June 24, 1778," *The Papers of George Washington*, Vol. 15, 528

[11] Lengel, ed., "War Council, June 24, 1778," *The Papers of George Washington*, Vol. 15, 521 and "General Washington to General Scott, June 24, 1778," *The Papers of George Washington*, Vol. 15, 534

[12] Lengel, ed., "General Orders, June 25, 1778," *The Papers of George Washington*, Vol. 15, 536

[13] Lengel, ed., "General Washington to General Lee, June 26, 1778," *The Papers of George Washington*, Vol. 15, 556

Battle of Monmouth

By the evening of June 27th, the American advance guard under General Lee was within striking distance of the British column at Monmouth Courthouse. What occurred the next day was one of the fiercest battles of the war; a twelve hour engagement in oppressive heat and humidity that tested the stamina and courage of everyone involved.

In spite of the caution of his war council, General Washington was determined to strike a blow upon the enemy so he ordered General Lee to attack as soon as the opportunity presented itself. Lee sent a detachment ahead of his advance guard at dawn with orders to make contact with the enemy and ordered the rest of his troops to proceed forward to support the detachment. Lee needed accurate information about General Clinton's movements, but unfortunately, conflicting reports from the militia made it difficult to determine whether the British army was on the march, or waiting for an attack at Monmouth Courthouse.

General Lee pressed forward and discovered that Clinton had resumed his march, leaving just a rear guard at Monmouth Courthouse. Lee was eager to envelope this force and maneuvered his several detachments to do so, but General Clinton undermined these efforts by sending a large portion of his army back to Monmouth to support his rear guard.

With his detachments spread out and difficult to coordinate and the enemy pouring forward towards the Americans, Lee's officers acted on their own to adjust their positions. This, unfortunately, created confusion among the advance guard as it appeared that detachments were retreating when they were really just re-deploying to a better fighting position. The confusion undermined General Lee's plan (which he had neglected to properly explain to his subordinates) and Lee was soon left with no choice but to abandon his attack and withdraw to a new defensive position to resist the unexpected British advance.

While all of this took place, General Woodford and his brigade were on the march, rushing forward with General Washington and the main body of the army to support Lee's advance guard. Before they reached Tennent's Meeting House, which was only a few miles west of Monmouth Courthouse, Washington ordered General Nathanael Greene to lead General Woodford and his brigade, along with four cannon, southeastwards towards Monmouth Courthouse via a secondary road.[14] Washington, who was unaware of Lee's retreat at the time he issued the order, wanted General Greene to screen the army's right flank while he marched with the rest of the army to support General Lee's advance guard at Monmouth Courthouse.[15]

While Greene and Woodford swept off to the right, General Washington rode ahead of the main body and discovered that his advance guard, which he believed was pushing the enemy, was actually withdrawing in disarray. Washington encountered General Lee on a small knoll on the road between Tennant Meeting House and Monmouth Courthouse and demanded to know why Lee's troops were retreating. Lee, taken aback by Washington's angry tone, stammered an

---

[14] Mark Edward Lender and Garry Wheeler Stone, *Fatal Sunday: George Washington, the Monmouth Campaign, and the Politics of Battle*, (University of Oklahoma Press, Norman, OK, 2016), 284

[15] David R. Hoth, ed., "General Washington to Henry Laurens, July 1, 1778," *The Papers of George Washington*, Vol, 16, (Charlottesville, VA: University of Virginia Press, 2006), 4

answer, which Washington impatiently dismissed. Washington then ordered Lee to the rear of the main body and acted to slow the enemy advance by reforming some of the advance guard himself and posting them with orders to hold their position as long as possible. He then rode back to the main body to hurry them into position along a ridge overlooking the area.

Brief, but fierce engagements erupted in the fields and orchards outside of Monmouth as pockets of Washington's advance guard challenged the British advance. They were too few, however, to stop the British alone, and by the afternoon the American advance guard had withdrawn across a creek and reformed behind the main American army, securely posted on Perrine Ridge. An intense artillery bombardment between the two sides ensued but neither gave any ground.

General Greene heard the gunfire on his left and adjusted his march, directing Woodford's brigade towards the fight to assist. Greene posted four cannon, covered by Woodford's brigade, upon a hill overlooking the left flank of the British line and commenced a deadly, enfilade fire. General Washington noted the importance of this fire in a letter to Congress.

Battle of Monmouth

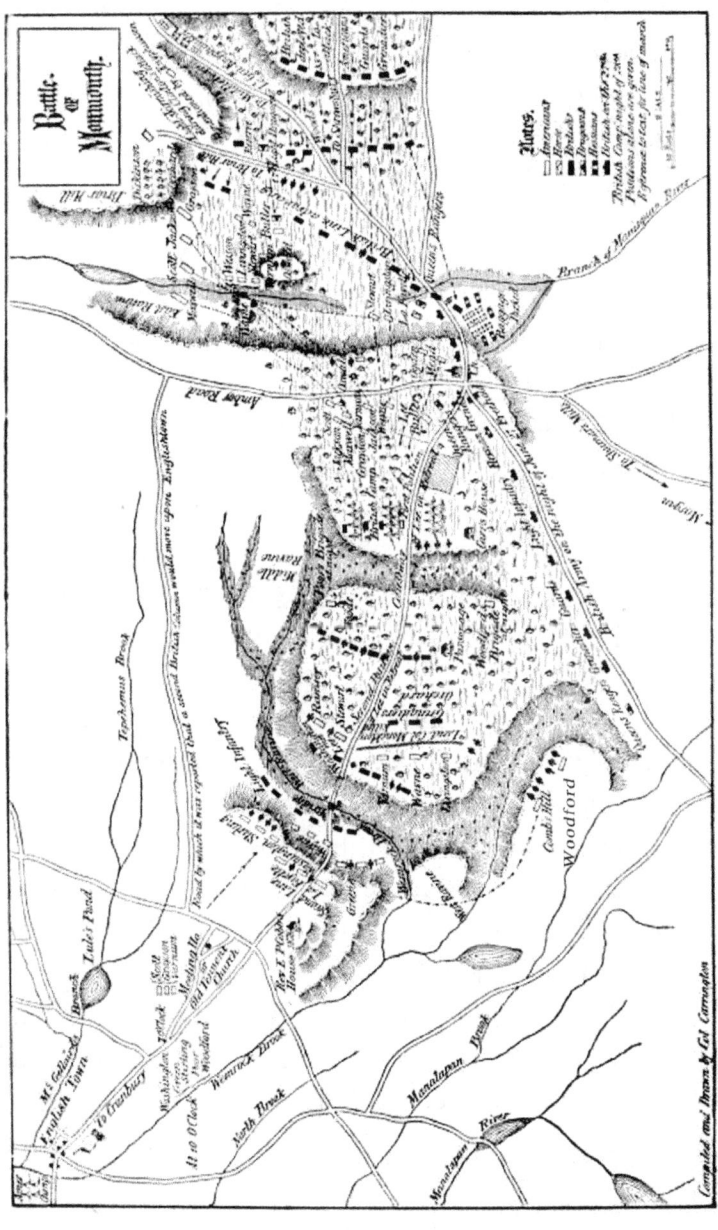

> *The Enemy...made a movement to our Right, with...little success, General Greene having advanced a Body of Troops with Artillery to a commanding piece of Ground, which not only disappointed their design of turning our Right, but severely enfiladed...*[the enemy's] *left Wing.*[16]

By the late afternoon, General Washington was pleased with the situation, the American army was holding firm on Perrine Ridge, but Washington wished to force the enemy back, so he ordered attacks upon their flanks. General Woodford's brigade attempted to attack the British left flank, but the swampy terrain at the bottom of the hill they were posted on and the fading light of day made an advance upon the enemy nearly impossible. The American troops attempting to strike the British right flank found similar obstacles, so General Washington rescinded his attack orders. General Woodford and his men spent a tense night on the hill expecting to renew the fight in the morning. To their surprise and likely their relief, General Clinton withdrew the British army after midnight, resuming his march from Monmouth to Sandy Hook and the waiting British fleet.

The American army was left to care for the wounded and bury the dead, a chore that General Woodford's brigade participated in by guarding the burial parties. Hundreds of troops on both sides perished in the battle, many succumbing to the heat with sunstroke. Although both sides held the field at the end of the day, General Washington viewed the British withdrawal as a sign of victory for his troops and

---

[16] Hoth, ed., "General Washington to Henry Laurens, July 1, 1778," *The Papers of George Washington*, Vol, 16, 4

congratulated the army in his orders.[17] He did not pursue General Clinton, who Washington believed was heading to New York, but instead, directed the army to march north in order to cross the Hudson River above New York and maintain pressure on the British.

At the same time, General Washington addressed an awkward situation that had developed with General Lee, namely, Lee's insistence that he had been mistreated by Washington when the two encountered each other outside Monmouth during Lee's retreat. Lee felt that General Washington had wronged him and requested a court martial to vindicate himself. General Washington promptly acquiesced and charged Lee with disobedience of orders for not effectively attacking the British, misbehavior before the enemy by making an unnecessary, disorderly, and shameful retreat, and disrespect to the Commander-in-Chief through his letters to General Washington after the battle.[18]

General Woodford was selected to serve on the twelve member court martial which met for almost six weeks before finding General Lee guilty on all counts and suspending him from the army for a year.[19] An argument could be made that justice eluded Lee, at least pertaining to the first two charges, but Lee's conduct towards Washington had doomed him. He returned to his estate in Virginia and died in 1782 of a fever while visiting Philadelphia.

Within two months of the Battle of Monmouth, General Washington positioned his army in the vicinity of White

---

[17] Lengel, ed., "General Orders, June 29, 1778," *The Papers of George Washington*, Vol. 15, 583

[18] Lengel, ed., "General Washington to General Lee, June 30, 1778, Footnote 2" *The Papers of George Washington*, Vol. 15, 597

[19] Ibid.

Plains, New York, just fifteen miles north of Manhattan Island and the British army. Washington's deployment to New York allowed him to unite with the smaller American northern army which was posted in the New York Highlands. This created issues over rank and command as officers from the two armies, which had operated largely independent from each other for nearly two years, were now consolidated under General Washington's direct command. General Washington addressed these issues of command and rank and also implemented changes to the composition of the Virginia brigades.

For General Woodford, Washington's new arrangement of the Virginia brigades meant a reunion with his original regiment, the 2nd Virginia, which was attached to Woodford's brigade in July.[20] Colonel Christian Febiger commanded the regiment, whose troop strength was bolstered when General Washington ordered the remnants of the 6th Virginia Regiment to merge with the 2nd Virginia.[21]

The consolidation and renumbering of several regiments within the Virginia line reduced the number of continental regiments from Virginia to twelve (from sixteen when Colonel Grayson's regiment is included). For General Woodford, this meant that his brigade of 900 effective officers and men consisted of only three "new" regiments (formed from six older and largely depleted regiments). Woodford's three new regiments were commanded by colonels Christian Febiger, William Heth, and Daniel Morgan, all outstanding officers.

---

[20] Hoth, ed., "General Orders, July 22, 1778," *The Papers of George Washington*, Vol. 16, 121
[21] Hoth, ed., "General Orders, August 4, 1778," *The Papers of George Washington*, Vol. 16, 241

Febiger commanded the combined 2nd and 6th regiments, Heth, the combined 3rd and 7th regiments, and Morgan, the combined 11th and 15th regiments.[22] If this wasn't confusing enough, although the 2nd and 3rd regiments kept their original designations, Woodford's old 7th Virginia was re-designated the 5th (and placed under Colonel Heth's command with the 3rd Virginia) and the old 11th Virginia became the 7th Virginia while the 15th Virginia became the new 11th Virginia, both of which fell under Colonel Morgan's command.[23]

The re-arrangement of regiments resulted in a surplus of officers without commands (supernumerary officers who were sent back to Virginia to recruit) and the elimination of one of Virginia's four brigades (Weedon's). The remaining three, Woodford's Muhlenberg's and Scotts, were temporarily placed under the command of Israel Putnam of Connecticut.[24]

In addition to serving on General Charles Lee's court martial and managing his own brigade, General Woodford attended several war councils at White Plains, advising Washington not to attack New York.[25] Washington accepted this advice, which was shared by all of the members of his war councils, and re-deployed most of the army approximately 40 miles north of White Plains in the vicinity of Fishkill and Fredericksburg, New York and Danbury, Connecticut.

---

[22] Lesser, "Monthly Returns of the Continental Army...Oct. 1, 1778," *The Sinews of Independence,* 89
[23] Ibid.
[24] Hoth, ed., "General Orders, September 7, 1778," *The Papers of George Washington*, Vol. 16, 534
[25] Hoth, ed., "War Council, July 25, 1778," *The Papers of George Washington*, Vol. 16, 160-163 and "General Woodford to General Washington, September, 2, 1778," 503

General Woodford and his brigade, along with the two other Virginia brigades of Putnam's division, marched to the east bank of the Hudson River, just a mile and a half south of West Point (which lay on the other side of the river) to support this crucial post.[26]

When British General Henry Clinton sent approximately 10,000 troops into the New York and New Jersey countryside in September to gather forage (4,000 north of White Plains and 6,000 into New Jersey) General Washington re-directed Woodford's brigade across the Hudson River towards the border of New York and New Jersey to hinder enemy foraging parties that strayed too far from their main position in Hackensack, New Jersey.[27]

General Washington sought to form a defensive line stretching in an arc from General Scott's position in Westchester, New York, northwestward to the main American army at Fredericksburg and Peekskill, across the Hudson River at West Point, then down to Clarkstown New York, ending at Elizabethtown, New Jersey where General William Maxwell's New Jersey brigade was posted.[28] Washington placed General Lord Stirling in command of the American troops in New Jersey and ordered him to, "cover the country," as best as he could. Maxwell's brigade was ordered north to the heights west of Acquaquenunk Bridge while Woodford

---

[26] Philander D. Chase, ed., "General Washington to Henry Laurens, September 23, 1778," *The Papers of George Washington*, Vol. 17, (Charlottesville, VA: University of Virginia Press, 2008), 93

[27] Todd W. Braisted, *Grand Forage, 1778: The Battleground Around New York City*, (Westholme: Yardley, PA, 2016), 71 and Chase, ed., "General Washington to General Israel Putnam, September 27, 1778," *The Papers of George Washington*, Vol. 17, 153

[28] Braisted, 115

rushed his brigade to Clarkstown.[29] Colonel George Baylor's 3rd Continental Light Dragoons, approximately 100 strong, was ordered to unite with Woodford's brigade at Clarkstown, but before they could do so, they were nearly annihilated by a British force that caught them by surprise early on the morning of September 28th.

Comprised mostly of Virginians, Colonel Baylor and his dragoons had spent several days in Paramus, New Jersey, just a few miles from the enemy, observing their activity. Feeling that he had pushed his luck by staying in one place for so long, Colonel Baylor moved his camp a few miles outside of Paramus on September 27th. In the pre-dawn hours of September 28th, a large body of British light infantry descended upon Baylor's sleeping men and nearly annihilated them. The vast majority of wounds inflicted upon the Americans came by British bayonets, many after the Americans had surrendered.[30]

General Woodford encountered the survivors of "Baylor's Massacre" the next day while on the march to Clarkstown and ordered those few who were still mounted to join him, the others proceeded on to General Washington's main army.[31]

Woodford's stay at Clarkstown was brief, General Washington desired that Woodford position his brigade to protect a strategic pass through the Ramapo Mountains (the

---

[29] Chase, ed., "Instructions to Major General Stirling, September 28, 1778," *The Papers of George Washington*, Vol. 17, 172
[30] Chase, ed., "Colonel Otho Holland Williams to General Washington, September 28, 1778,", 173-174 and
"David Griffith to General Stirling, October 20, 1778," in note 1, 457 *The Papers of George Washington*, Vol. 17
[31] Chase, ed., "General Woodford to General Washington, September 29, 1778," *The Papers of George Washington*, Vol. 17, 197

Clove) that served as a back door to West Point. As a result, Woodford marched southwestward a few miles to Paramus, New Jersey and encamped there.[32]

Significantly outnumbered by the enemy, there was little General Woodford or any of the Americans in New Jersey could do to halt the extensive foraging of General Clinton's troops. Woodford informed Washington that the British,

> *Appear to be busy at work upon two Redoubts on this side of the new Bridge, & their Forageing partys for the other side are very Strong – I keep out small scouting partys for the purpose of gaining intelligence, but our numbers will not afford one large enough to cope with those of the Enemy, who are never out of supporting distance of a Battalion or two of Light Infantry.*[33]

General Woodford added that recently captured British deserters claimed that General Clinton intended to send ten or more regiments to the West Indies very shortly to defend those valuable islands from the French.[34]

The British remained in New Jersey for another week, collecting as much forage as possible against minimal American resistance. General Woodford informed Washington on October 13th of their departure.

---

[32] Chase, ed., "General Washington to General Woodford, September 30, 1778," *The Papers of George Washington*, Vol. 17, 212-213 and "Major General Stirling to General Washington, October 1, 1778, 217

[33] Chase, ed., "General Woodford to General Washington, October 4, 1778," *The Papers of George Washington*, Vol. 17, 262

[34] Ibid.

> *The Enemys Rear left the New Bridge this morning after setting Fire to their Redoubts & Hutts, they took with them several of the Inhabitants, some by force, & others voluntarily went with them. I have had partys of Horse round them all Day....* [35]

Woodford added that more British deserters confirmed the earlier reports that General Clinton intended to send ten or more regiments to the West Indies as soon as the foraging expedition was complete.

The withdrawal of the British prompted General Stirling to order Woodford's brigade to take post in Newark. Stirling informed Washington of his decision and asked that Woodford's baggage and artillery be forwarded from camp to Newark.[36] General Washington replied that he did not intend to keep Woodford's brigade in Newark much longer, so he would send only the baggage that was absolutely necessary as well as the requested cannon.[37]

As it turned out, Woodford's brigade did remain in New Jersey for the rest of the year and through the winter, but it did so without its commander. With the campaign season apparently finished, General Woodford was granted a furlough to return to Virginia to settle the affairs of his deceased father-in-law. Woodford departed on October 21st, leaving command of his brigade to Colonel Daniel Morgan.[38]

---

[35] Chase, ed., "General Woodford to General Washington, October 13, 1778," *The Papers of George Washington*, Vol. 17, 371-372

[36] Chase, ed., "General Stirling to General Washington, October 14, 1778," *The Papers of George Washington*, Vol. 17, 380

[37] Chase, ed., "General Washington to Major General Stirling, October 17, 1778," *The Papers of George Washington*, Vol. 17, 429

[38] Chase, ed., "General Washington to Samuel Washington, October 22, 1778," *The Papers of George Washington*, Vol. 17, 536

Chapter Ten

"For my Part I would Wish Never to Part with Him."

1778-1779

General Woodford likely reached Windsor, his home in Caroline County, before the start of November. The Virginia Assembly had already acted on behalf of his wife and the other heirs of Woodford's father-in-law, John Thornton, who had died in June with no will. The legislature agreed to an application made by General Woodford and the husbands and fathers of Thornton's heirs (John Taliaferro, Samuel Washington, and John Lewis) that trustees be named to manage the sale of Thornton's property in order that the heirs receive the best value they could for their inheritance.[1] The idea was that Thornton's several large properties would be sold intact, thus deriving the highest price possible for them. The proceeds of the sales would then be used to purchase property more equitable for the several heirs so the value of Thornton's estate could be divided evenly.

While General Woodford focused on this business, weekly accounts in the Virginia gazettes of the failed British efforts to restore the colonies to the empire dominated the news.[2] The

---

[1] Henings, ed., "October 1778," *The Statutes at Large Being a Collection of all the Laws of Virginia*, Vol. 9, 573-574

[2] Dixon and Hunter, "October 30, November 6, 13, 27, December 4, 1778,"

response of American writers to Britain's peace commission and other overtures was, too little, too late. Americans were resolved for independence, and with France and other European countries now entering the dispute on their behalf, Americans were more confident than ever that they would soon secure their independence.[3] Expressions of confidence in the cause were particularly needed at that time because the situation for many Virginians (as well as most Americans) was one of economic hardship caused by the long disruption of trade and the significant depreciation of money brought on by the decision of Congress and the states to print large amounts of paper currency.[4]

As with General Woodford's earlier furloughs, we know few details of how he spent his time in Virginia. He likely visited family and friends, especially during the Christmastide season while waiting to settle his father-in-law's estate. The delay in settling this matter forced General Woodford to sheepishly write General Washington on January 3rd, 1779 with an explanation for his delayed return to camp in New Jersey:

Virginia Gazette, and Dixon and Nicolson, "February 12, 19, 26, and March 5, 12, 1779," *Virginia Gazette*

[3] Dixon and Hunter, "The Crisis No. 6, November 27, 1778," *Virginia Gazette*, 1

[4] Dixon and Nicolson, "Philadelphia, In Congress, January 13, 1779," and "Williamsburg, From the United States Magazine, February 19, 1779," *Virginia Gazette*, February 19, 1779, 2-3 and
Edward G. Lengel, ed., "Lt. Col. Holt Richeson to General Washington, April 14, 1779," *The Papers of George Washington*, Vol. 20, (Charlottesville, VA: University of Virginia Press, 2010), 70-71

When I obtain'd your Excellencys permission to come to Virginia I had no doubt but the business that brought me in, would by this time have been completed, & that I should have been able to comply with your Excellencys request, & my own inclinations, to return to my Duty in the army – but it will necessarily detain me some considerable time longer; for which I hope I shall have your further indulgence, when I assure you that nothing but an Affair of the greatest consequence to my Family could induce me to take the liberty to Trespass upon your Excellencys goodness in permitting me to leave the Army so early in the Fall – I will not presume upon your time by entering into a particular detail of the business that I am employd in, but only to Assure you I will use the greatest industry in my power to put Affairs into such a way as will enable me to Join the Army   I fear it will not be possible to do so till the last of Feby.

A line from your Excellency, or one of the Gentlemen of the Family will relieve me from great anxiety – as there is nothing I wish more to avoid then incurring your disapprobation of my conduct as an Officer -- & on the other Hand, my peculiar situation, as to my private concerns, demand my attention at least for the time I have mention'd.[5]

During Woodford's absence from the army, command of his brigade, which was encamped in winter huts in Middlebrook, New Jersey, fell to Colonel Daniel Morgan. Colonel Morgan had returned to the brigade a month after the

---

[5] Edward G. Lengel, ed., "General Woodford to General Washington, January 3, 1779," *The Papers of George Washington*, Vol. 18, (Charlottesville, VA: University of Virginia Press, 2008), 562-563

battle of Monmouth to command his original regiment (the 11th which was re-designated the 7th Virginia in the fall) because his vaunted rifle corps had shrunk to only 100 men. Over the winter Woodford's brigade, like much of the army, also shrunk to about 550 effective officers and men by the end of 1778.[6]

More than a month after Woodford penned his early January letter to General Washington, the American commander-in-chief, who had yet to receive Woodford's letter, impatiently wrote to Woodford to urge him to return to camp immediately.

> *The Circumstances and Situation of the Virginia line call loudly for your return to the Army as soon as possible, more especially as General Muhlenberg, the only General Officer of the state now present, has long had a promise of leave to visit his family and private Affairs whenever he could be possibly spared. As the time which you expected to be absent has considerably elapsed, I am not without hopes that you will be here before this reaches Virginia. Should it find you there, I must desire you to set out for the Army immediately upon the rect of it.*[7]

Woodford received Washington's letter in late February and responded with an explanation for his continued delay.

---

[6] Lesser, "Monthly Return of the Continental Army...for January 1779," *The Sinews of Independence*, 100

[7] Philander D. Chase and William M. Ferraro, eds., "General Washington to General Woodford, February 10, 1779," *The Papers of George Washington*, Vol. 19, (Charlottesville, VA: University of Virginia Press, 2009), 171

I recd your favor of the 10th Feby only three Days ago. I should have set out immediately for Camp as your Excellency desired, but am under an obligation to attend Caroline Court, which happens the 11th Day of the month, to settle my Administration acct of Colo. Thorntons Estate, this could not be done at the Feby Court, owing to the badness of the weather, & will lay me under the disagreeable necessity of risqueing your Excellencys displeasure, then which, nothing could give me more real uneasiness. I am not without hope that I shall be able to satisfy your Excellency that nothing but the most pressing necessity could have induced me to remain so long here.

I am extremely sorry that my stay should have detain'd Genl Muhlenberg or any other Officer from visiting their Family or private Affairs; I was apprehensive some time ago that this might be the case & strove to avoid it by every means in my Power.

Your Excellency may be assured that I will set off the Day after the Court, & use the greatest expedition in getting to Camp.[8]

While he waited for the Caroline County Court to convene in March, Woodford completed the sale of his portion of the brewery in Fredericksburg and arranged for the sale of twenty slaves (probably John Thornton's).[9] His advertisement announcing the sale was typical of such ads found in the Virginia gazettes.

---

[8] Chase and Ferraro, eds., "General Woodford to General Washington, March 1, 1779," *The Papers of George Washington*, Vol. 19, 318

[9] Stewart, *The Life of Brigadier General William Woodford of the American Revolution*, Vol. 2, 1013

To be sold to the highest bidder for ready money, before the coffeehouse door in Fredericksburg, on Thursday the 21st of February (being Spotsylvania court day) TWENTY VALUABLE NEGROES.[10]

True to his word, General Woodford departed for New Jersey to rejoin his brigade upon the completion of his affairs at the Caroline County court. His eldest son, John, accompanied him to Princeton where he enrolled in Dr. John Witherspoon's fine College of New Jersey.[11]

General Woodford was back in camp at Middlebrook by early April, commanding his brigade and participating in meetings of a general officers board, ironically to settle disputes over rank among various officers of the army.[12] His brigade had increased to over 700 effective officers and men, which followed the typical pattern of growth for the American army that occurred every spring when many of the troops returned from furlough.[13]

Reports that General Clinton had ordered nine British regiments aboard ships in New York, presumably to reinforce British forces in either Georgia or the West Indies, briefly presented General Washington with an opportunity to attack New York in conjunction with the French, but Count D'Estang, commander of French naval forces in America, would

---

[10] Dixon and Nicolson, "Slave Ad, February 12, 1779," *Virginia Gazette*, 4
[11] Stewart, 1014
[12] Lengel, ed., "General Orders, April 8, 1779," *The Papers of George Washington*, Vol. 20, 1 and "Board of General Officers to General Washington, April 13, 1779," 49
[13] Lesser, "Monthly Troop Returns of the Continental Army for April 1779," 112

not risk his warships or troops.[14] Washington instead decided to reinforce the small American Southern Army in Charleston, South Carolina under General Benjamin Lincoln with new Virginia recruits raised by General Charles Scott.

General Scott had returned to Virginia the previous November to restore his health and had remained there into the spring to recruit and organize new levies meant to reinforce the depleted Virginia regiments with Washington's army. On May 5th, General Washington ordered General Scott to organize the new Virginia recruits (over a thousand strong) into three battalions and then lead them as a brigade to South Carolina to reinforce General Lincoln.[15] This order did not initially apply to officers and men who were in Virginia on furlough because they were already members of Virginia regiments in New Jersey. Scott was instructed to order the troops on furlough to return north as quickly as possible, but just a few days later General Washington decided to allow Scott to keep those troops who were willing to march south with him instead of returning to camp in New Jersey.[16]

With General Scott re-assigned to the South, General Washington dissolved Scott's brigade, placing the 4th and 8th Virginia Regiments into Woodford's brigade, and the remainder of Scott's troops into General Muhlenberg's

---

[14] Lengel, ed., "General Washington to Conrad-Alexandre Gerard, May 1, 1779," *The Papers of George Washington*, Vol. 20, 279-280
[15] Lengel, ed., "General Washington to General Scott, May 5, 1779," *The Papers of George Washington*, Vol. 20, 342-343
[16] Lengel, ed., "General Washington to General Scott, May 5, 1779," *The Papers of George Washington*, Vol. 20, 342-343 and "General Washington to General Scott, May 12, 1779," *The Papers of George Washington*, Vol. 20, 456-457

brigade.[17] Officers who found themselves without troops to command (supernumerary) were sent back to Virginia to recruit, several opting instead to join General Scott's expedition southward.

The addition of two more regiments, as well as the return of furloughed troops and a few new recruits, increased General Woodford's effective force of officers and men to over 1,200 by the end of May.[18] Less than a month later, General Washington re-organized the Virginia brigades (as well as those from Maryland and Pennsylvania) again, merging the 2nd Virginia with the 5th and 11th regiments to form one battalion, the 7th regiment with the 8th regiment to form another battalion, and the 3rd and 4th regiments to form a third battalion for Woodford's brigade.[19] Each new battalion averaged over 350 effective officers and men, which was still a far cry from full strength, but allowed Washington to send even more officers to Virginia to recruit reinforcements.[20]

General Washington also moved to create a replacement unit for Colonel Daniel Morgan's rifle corps. Washington opted for a corps of light infantry, based largely on the British model of light infantry. He ordered the newly merged battalions of the Virginia, Maryland and Pennsylvania troops to select a number of their best soldiers to serve in light

---

[17] Lengel, ed., "General Orders, May 12, 1779," *The Papers of George Washington*, Vol. 20, 444-445

[18] Lesser, "Monthly Troop Return of the Continental Army, May 1779," 116

[19] William M. Ferraro, ed., "General Orders, June 12, 1779," *The Papers of George Washington*, Vol. 21, (Charlottesville, VA: University of Virginia Press, 2012), 138-139

[20] Lesser, "Monthly Troop Return of the Continental Army, June 1779," 120

infantry companies.[21] Once the active campaign season began, these companies would be temporarily detached from their original battalions and formed into a light infantry battalion. This new unit would serve several functions for General Washington including reconnaissance and perimeter defense as well as shock troops when the situation called for quick and decisive action.

At the end of May, General Washington asked General Woodford and his other general officers for their opinions on how to proceed with the upcoming campaign season. He had already decided to send General Sullivan with several brigades into western Pennsylvania and New York to subdue the Indians and Tories that had long plagued the region. The commander-in-chief desired to know what his generals thought he should do with the rest of the army, act offensively or defensively.[22]

Washington's inquiry became mute just days after he broached the subject when General Clinton sent a large British force up the Hudson River to seize Stony Point and Verplanck's Point, eliminating Kings Ferry as a continental crossing spot and threatening the American post at West Point, just fifteen miles upriver. General Washington immediately ordered the army, which was still encamped in New Jersey, New York, and Connecticut, to move into better positions to defend West Point. General Woodford and his brigade marched north into New York, camping just a few miles

---

[21] Ferraro, ed., "General Orders, June 12, 1779," *The Papers of George Washington*, Vol. 21, 138-139

[22] Lengel, ed., "Circular to General Officers, May 28, 1779," *The Papers of George Washington*, Vol. 20, 651-652

southwest of West Point.[23] He penned a letter to his son John in Princeton on June 13th, forwarding him a letter from his brother Catesby and updating him on both home and military affairs. One senses all of the universal concerns of parents in the letter as well as the affection General Woodford holds for his eldest son.

Dear Jack,
I received from your mama yesterday a letter inclosing one from our Dear little Catesby for you which you will receive under this cover.
The family at Windsor are all well except Mr. Saunderson who has been dangerously ill for sometime. Your mama complains of your writing her but seldom and also of your want of attention to your manner of writing and spelling. I will hope that you will give so good a mother no future reason to complain on either account; not only for her satisfaction and pleasure, but for your own improvement. If you apply to the quartermaster at Prince Town he will forward your letters to me by express coming frequently to Gen. Greene, and I hope to hear from you often. We have been five or six days in this place watching the Enemies motion who are fortifying at King's Ferry, on both sides of the North River; we are about twelve miles from them and the same distance from our important post at West Point, which we suppose their object. I will write to you again when anything interesting, and shall rest assured that you will make the closest application to your studys,

---

[23] Ferraro, ed., "General Orders, June 12, 1779," *The Papers of George Washington*, Vol. 21, 93

and cultivate an acquaintance with the Genteelest Families in your neighborhood; it will give me pleasure to hear that you are frequently at Gen. Morris's.
Present the Gentlemen and Ladies of Prince Town with my respectful compliments and be assured my Dear Boy, that I am very truly,
Your affectionate Father,
Wm. Woodford [24]

By late June, General Woodford was in New Windsor, several miles north of West Point, where he served on a Board of General Officers to determine the status of American prisoners paroled by or who had escaped from British captivity (in possible violation of their paroles).[25] General Woodford and his fellow officers were able to address such matters because the British showed no sign of advancing any further up the Hudson River; they seemed content to fortify their new positions at Stony Point and Verplanck's Point. General Washington, however, was not content to leave them undisturbed.

In mid-July Washington's new light infantry corps staged a daring night attack upon Stony Point that resulted in the capture of over 500 British troops. Although General Woodford and his brigade did not participate in this attack, a number of Woodford's officers and men, attached to the light infantry, did. News of their success raced through the country and boosted American morale.

---

[24] Stewart, "William Woodford to John Woodford, June 13, 1779, *The Life of Brigadier General William Woodford of the American Revolution*, Vol. 2, 1043

[25] Ferraro, ed., "Board of General Officers to General Washington, June 28, 1779," *The Papers of George Washington*, Vol. 21, 279-282

General Anthony Wayne, who led the attack, realized that it was impossible for his troops to keep Stony Point, so the light corps withdrew northward with their prisoners. Unsure whether the British would retaliate with a strike against West Point, General Washington kept his army on alert and hastened efforts to fortify the post. He ordered General Woodford and his fellow Virginians, along with General Wayne's light infantry corps, to guard the approaches to West Point, particularly those to the west.[26] By the end of the month Woodford and Muhlenberg's brigades were encamped on the border of New York and New Jersey near Suffern, New York.[27] Although he wasn't certain, it began to appear to General Washington that the British had no intention of attacking West Point, so Washington pondered where he might be able to attack them.

Powles Hook

Powles Hook sat on a peninsula that jutted into the Hudson River, within sight of New York City and the British fleet anchored in the harbor. The 400 man garrison at Powles Hook was protected on three sides by the Hudson River and from the west by a large marsh that flooded at high tide.[28] The only land approach to the fort was over a long causeway through the marsh and over a drawbridge that spanned a tidal moat dug across the flat peninsula. A ring of abattis (piled up brush designed to slow enemy attacks) encircled the post which

---

[26] Ferraro, ed., "General Orders, July, 19, 1779," *The Papers of George Washington*, Vol. 21, 567
[27] Ferraro, ed., "General Washington to Major General Stirling, July 24, 1779," *The Papers of George Washington*, Vol. 21, 639
[28] John W. Hartmann, *The American Partisan: Henry Lee and the Struggle for Independence: 1776-1780*, (Burd St. Press, 2000), 106-107

included two fortified redoubts bristling with cannon and two block houses.[29]

Although General Washington was initially skeptical of a strike against Powles Hook, Major Harry Lee of Virginia convinced both Washington and General Stirling (whose troops would participate in the attack) of its feasibility. Lee, just 23 years old, had developed a well deserved reputation as a daring cavalry commander and had become a favorite of Washington, no doubt in part because Washington was well acquainted with Lee's father in Virginia (they had served together in the Virginia House of Burgesses).

On August 18th, Major Lee led a detachment of 350 troops (that included men from General Woodford's brigade) from General Stirling's headquarters in Hackensack, New Jersey to Powles Hook, some eighteen miles to the south. Lee hoped to time the long march so that his attack commenced at half past midnight, a few hours before high tide. Unfortunately, Lee's guide on the march went astray and several crucial hours were lost marching through difficult terrain.

When they finally arrived to the edge of the marsh they were three hours late and without a quarter of their force, who had straggled behind. Informed by a scout that all was quiet in the fort and the marsh and moat that defended it was still passable despite the rising tide, Lee decided to attack immediately. He explained his decision in a letter to General Washington.

---

[29] Ibid. 107

I found my [original plan of attack] *impracticable, both from the near approach of day, and the rising of the tide. Not a moment being to spare, I paid no attention to the punctilios of honor or rank, but ordered the troops to advance in their then disposition.*[30]

Lee pushed his men forward, determined, *"to leave my corpse within the enemy's line,"* if the attack failed.[31] Captain Levin Hardy commanded a company of Maryland troops in the assault and described the advance.

We had a morass to pass of upwards two miles, the greatest part of which we were obliged to pass by files, and several canals to ford up to our breast in water. We advanced with bayonets, pans open, cocks fallen, to prevent any fire from our side....[32]

Lee's men rushed the outer works and stormed into the fort. A bit of fortune shined on them when they discovered that the main gate was open in expectation of the return of a large Tory patrol. This also meant that the garrison was weaker than anticipated. The Americans poured into the fort, but they still had two strong redoubts and two fortified block houses to overcome.

---

[30] Frank Moore, ed., "Extract of a letter from an officer at Paramus," *Diary of the American Revolution*, Vol. 2, 207
[31] William B. Reed, "Henry Lee to President Reed, 27 August, 1779," *Life and Correspondence of Joseph Reed*, Vol. 2, (Lindsay and Blakiston: Philadelphia, 1847), 126-27
[32] Reed, "Levin Handy to George Handy, 22 July, 1779," *Life and Correspondence of Joseph Reed*, Vol. 2, 126

Lieutenant McAllister, supported by Major John Clark of General Woodford's brigade, led troops that easily overwhelmed the dazed defenders of the center redoubt, capturing six cannon and the post's colors.[33] At the same time, another detachment captured one of the block houses.[34] The main barracks of the post, filled with invalids and camp followers, also fell quickly, but the fort's commander, Major Nicholas Sutherland and about 25 Hessians successfully defended another redoubt.[35]

Up to this point the Americans had not fired their muskets, relying instead on surprise and their bayonets to subdue the enemy. Most were unable to fire anyway as their gunpowder had gotten wet fording the canals and moat. With alarm guns firing across the Hudson River and the British army and navy rousing itself into action, it was imperative that Lee begin his retreat to safety with his 150 prisoners.[36] Lee recalled that,

The appearance of daylight, my apprehension lest some accident might have befallen the boats [that Lee and the prisoners were to march to] *the numerous difficulties of the retreat, the harassed state of the troops, and the destruction of all our ammunition by passing the canal conspired in influencing me to retire at the moment of victory.*[37]

---

[33] Hartmann, 114

[34] Mark M. Boatner, *Encyclopedia of the American Revolution*, (NY: D. McKay Co., 1966), 839

[35] Hartmann, 114-115

[36] Reed, "Levin Handy to George Handy, 22 July, 1779," *Life and Correspondence of Joseph Reed*, Vol. 2, 126

[37] Moore, ed., "Extract of a letter from an officer at Paramus," *Diary of the American Revolution*, Vol. 2, 208

Major Lee halted the assault on the remaining enemy redoubt and ordered his detachment to march westward with their prisoners, sparing the fort's barracks from destruction because it was occupied by women and children and sick soldiers.[38] Lee also failed to spike the enemy cannon. There was just no more time left.

To reduce the chance of being intercepted by a large enemy force from New York, Major Lee followed a different route westward for the return march, one that required the detachment to cross the nearby Hackensack River by boat. Unfortunately, while on the march to the boats, Lee learned that Captain Henry Peyton's cavalry detail, tasked with guarding the boats, never received word of Lee's delay and had departed with the boats at sunrise, assuming that Lee had cancelled the attack.[39] With no way across the Hackensack River, Lee was forced to retrace his march northward along his original route and risk interception by the enemy. He recalled,

> *In this very critical situation, I lost no time in my decision, but ordered the troops to regain Bergen road.... Oppressed by every possible misfortune, at the head of troops worn down by a rapid march of thirty miles, through mountains, swamps, and deep morasses, without the least refreshment during the whole march, ammunition destroyed, encumbered with prisoners, and a retreat of fourteen miles to make good, on a route admissible of interception at several points...one* [enemy] *party moving in our*

---

[38] Ibid., 212
[39] Ibid.

> *rear and another...in all probability well advanced on our right, a retreat naturally impossible to our left, under all these distressing circumstances, my sole dependence was in the persevering gallantry of the officers, and obstinate courage of the troops.*[40]

Once again, fortune shined on Lee and his men when they encountered a detachment of fifty Virginians (likely part of Lee's missing men) with dry gunpowder.[41] Lee halted long enough to distribute a few cartridges to each man and then continued on. When they reached the vicinity of Fort Lee they were met by another large body of troops, reinforcements sent by General Stirling. Lee's men could now effectively defend themselves, which is what they did when an enemy detachment suddenly emerged on their flank. After a brief skirmish, the Americans pushed on to New Bridge and safety.

Major Lee's exhausted detachment reached camp around 1:00 p.m., after nearly twenty-four hours of constant activity. They had marched nearly 20 miles under difficult conditions to surprise the enemy at Powles Hook. At the loss of just a handful of men, they captured over 150 enemy troops and killed or wounded another 50.[42] They then marched another 20 miles past an alarmed enemy, burdened with enemy prisoners that they guarded with virtually empty muskets.

Praise for Lee and his expedition was extensive and included General Washington, who expressed his gratitude to Lee and his men in the general orders.

---

[40] Ibid., 209
[41] Ibid.
[42] Reed, "Levin Handy to George Handy, 22 July, 1779," *Life and Correspondence of Joseph Reed*, Vol. 2, 126

> *The General has the pleasure to inform the army that on the night of the 18th instant, Major Lee at the head of a party composed of his own Corps, and detachments from the Virginia and Maryland lines, surprised the Garrison of Powles Hook and brought off a considerable number of Prisoners with very little loss on our side. The Enterprise was executed with a distinguished degree of Address, Activity and Bravery and does great honor to Major Lee and to all the officers and men under his command, who are requested to accept the General's warmest thanks.*[43]

Not everyone, however was pleased with Lee's conduct. A group of Virginia officers in General Woodford's and General Muhlenberg's brigades accused Lee of misconduct in the attack. They charged Lee with lying about the date of his commission to deny Major Clark, who actually outranked Lee, his rightful place of command of the detachment. They also accused Lee of leading a disorderly attack and an unnecessary and disorderly retreat from Paulus Hook. General Woodford and General Muhlenberg shared these accusations with General Washington, who was more troubled by the fact that they had been brought forth than he was that they might be true. Washington tried to resolve the dispute quickly and quietly, but the accusers, who may have included Woodford, insisted that Lee face a court martial, which he did in September.

---

[43] John C. Fitzgerald, ed., "General Orders, 22 August, 1779," *The Writings of George Washington*, Vol. 16, (Washington, D.C.: U.S Government Printing Office, 1937), 149

Major Lee was stung by the accusations, especially as they came from his own countrymen (fellow Virginians). Lee shared his troubles with his friend, Joseph Reed, President of Pennsylvania's Supreme Executive Council.

> *Generals and Colonels are now barking at me with open mouth. Colonel Gist, of Virginia, an Indian hunter, has formed a cabal. I mean to take the matter very serious, because a full explanation will recoil on my foes, and give new light to the enterprise.*[44]

Lee confessed that he had actually been too generous in his praise for some of the troops in the attack.

> *I did not tell the world that near one half of my countrymen (fellow Virginians) left me – that it was reported to me by Major Clarke as I was entering the marsh, -- that notwithstanding this and every other dumb sign, I pushed on to the attack.*[45]

Lee asserted that he had been prepared to sacrifice his life in the attempt on Powles Hook while the efforts of many of Major Clark's Virginians were, "not the most vigorous." Lee ended the letter by assuring Reed that

> *I am determined to push Colonel Gist and party. The brave and generous throughout the whole army support me warmly..... I have received the thanks of General Washington in the most flattering terms, and the congratulations of General Greene* [and] *Wayne.*

---

[44] Reed, "Henry Lee to President Reed, 27 August, 1779," *Life and Correspondence of Joseph Reed*, Vol. 2, 126
[45] Ibid.

Do not let any whispers affect you, my dear sir. Be assured that the more full the scrutiny, the more honour your friend will receive and the more ignominy will be the fate of my foes.[46]

The charges against Major Lee at his court martial centered on his illegitimate command of the detachment (he was technically outranked by Major Clark) and his conduct of the attack and retreat (both of which were described as disorderly).[47]

General Washington clearly wished the issue to be settled in Lee's favor and provided evidence in the form of a letter to discredit the charge that Lee's retreat was too hasty.[48] Even Major Clark, (who was denied the honor of command because Lee allegedly lied about the date of his commission), helped Major Lee by testifying on his behalf [49] After five days of testimony the court rendered its decision on September 11th. Describing some of the charges against Lee as, "unsupported" and "groundless", and some of his actions as necessary and fully justified, the tribunal acquitted Major Lee with honor on all eight charges.[50]

The extent of General Woodford's role in Lee's court martial is unclear. A letter from Edmund Pendleton, suggests that Woodford may not have been one of Lee's main accusers.

---

[46] Ibid., 127
[47] Fitzgerald, ed., "General Orders, 11 September, 1779," *The Writings of George Washington*, Vol. 16, 262-265
[48] Fitzgerald, ed., "General Washington to Major Henry Lee, 1 September, 1779," *The Writings of George Washington*, Vol. 16, 217-218
[49] Hartmann, 123
[50] Fitzgerald, ed., "General Orders, 11 September, 1779," *The Writings of George Washington*, Vol. 16, 262-265

> *I can't help being pleased to hear of Major Lee's acquittal, as I suppose him a good Officer, tho' have reason to believe he has been too highly puffed by some Family Partizans. I hope Our Countrymen his Accusers have not Suffered in their reputation by his Acquittal, further than having been mistaken in his supposed guilt.*[51]

It is unlikely that Pendleton would have expressed this hope if Woodford had indeed been one of Lee's accusers. Then again, Pendleton's declaration could have been a subtle rebuke of Woodford. It is difficult to say.

Stalemate in the North

A week prior to Major Lee's acquittal, General Woodford wrote to Pendleton thanking him for the news that Woodford's brother, Captain Henry Woodford, a mariner who had the misfortune of being captured by the British navy in the Mediterranean, had been released by the British and had returned to Virginia.

> *I am much obliged by your congratulations upon my Brother's safe Return. I expect the Capt. has paid you a visit before now. He has been particularly unfortunate, and I believe will be glad to get into the way to be revenged.*[52]

---

[51] Mays, ed., "Edmund Pendleton to William Woodford, October 11, 1779," *The Letters and Papers of Edmund Pendleton*, Vol. 1, 300

[52] Stewart, "William Woodford to Edmund Pendleton, September 2, 1779," *The Life of Brigadier General William Woodford of the American Revolution*, Vol. 2, 1082-1083

Woodford added that a British reinforcement of a couple of thousand new recruits had recently arrived in New York, but there was little concern about this in the American camp.[53]

Despite enduring two surprise American attacks on outposts considered safe and secure, attacks that cost the British army hundreds of soldiers, General Clinton had done nothing to retaliate. He seemed content to sit in New York City and wait for winter to arrive.

While General Washington briefly considered an attack on New York in coordination with the French (something the French never seriously contemplated) Woodford and his Virginians remained encamped a few miles to the southwest of West Point, guarding against a possible British surprise. In mid-October Woodford assumed command of General Stirling's division (the Virginia continentals) while Stirling left the army to recover his health.[54] In truth, General Woodford's health was not ideal; he informed General Stirling a few days earlier that he had been confined to his quarters for two days with a fever and lax.[55] Woodford soon recovered and took command of the division, corresponding and dining frequently with General Anthony Wayne, whose light infantry corps was also posted in the area.

One of General Woodford battalion commanders, Colonel John Nevell, shared with Colonel Daniel Morgan (who was home in Virginia) an account of the brigade's activities in the

---

[53] Ibid.
[54] Benjamin L. Huggins, ed., "General Washington to General Woodford, October 12, 1779," *The Papers of George Washington*, Vol. 22, (Charlottesville, VA: University of Virginia Press, 2013), 717-718
[55] Stewart, "General Woodford to General Stirling, October 8, 1779, *The Life of Brigadier General William Woodford of the American Revolution*, Vol. 2, 1092

fall. Morgan had left the army in disgust in July after General Anthony Wayne was selected instead of Morgan to command the new light infantry corps.

> *We have been imployed in making fascines and gabions upon which the Enemy left Stony Point and Planks Point, we are rebuilding a small part of it which must be a post for some poor field officer this winter....General Woodford has had the Command of the Division for some time Past and am Sorry to inform you he is very Much Disliked in Perticular by his Old Brigade Much more then by those that joined him this Campaign for my Part I would wish never to Part with him & the Old Lord (Sterling). However officers are willing to a man....*[56]

It is difficult to determine why so many of Woodford's officers disliked him so much, but one possibility was his long held belief in strict military discipline. Colonel Morgan, who had served under Woodford for over two years (excluding the period Morgan was detached with the rifle corps) appeared to hold Woodford in high regard, at least based on the letters he wrote to Woodford while he was home in Virginia.[57] There was no doubt, however, that the personalities of both men were very different. Morgan, whose humble background made him extremely comfortable among the men, was likely one of the most popular officers in the army. Woodford, who

---

[56] Stewart, "Colonel John Nevell to Colonel Daniel Morgan, November 9, 1779, *The Life of Brigadier General William Woodford of the American Revolution*, Vol. 2, 1114-1115

[57] Stewart, "Daniel Morgan to General Woodford, October 3, 1779, *The Life of Brigadier General William Woodford of the American Revolution*, Vol. 2, 1090

according to one of his sergeants was, "*the Damndest Partial Rascal on this earth without exception*," was obviously not.[58]

General Washington was another officer who held General Woodford in high regard, and that was all that really mattered to Woodford. Washington spent much of October at West Point hoping the French might cooperate in an attack on New York, but by the end of the month he grudgingly realized the French weren't interested and abandoned the idea. Frustrated by his inability to act and confused by the inactivity of the enemy, Washington confessed in a letter to his friend and fellow Virginian, Benjamin Harrison in late October that he was thoroughly perplexed on how to proceed against the British. The actions of General Clinton baffled Washington.

The enemy have wasted another Campaign...There is something so truely unaccountable in all this that I do not know how to reconcile it with their own views, or to any principle of common sense.[59]

General Washington did not understand why General Clinton had taken the trouble to establish two strong outposts at Stony Point and Verplancks Point in June, only to abandon them a few months later at the loss of hundreds of troops. Clinton had also done nothing to assist his Indian allies against a summer expedition that General Washington had launched in western Pennsylvania and New York. Clinton's brief naval raid on Virginia in May, although destructive and disruptive to Virginians in southeastern Virginia, seemed rather arbitrary

---

[58] Ward, 123

[59] William M. Ferraro, ed., "General Washington to Benjamin Harrison, October 25, 1779," *The Papers of George Washington*, Vol. 23, (Charlottesville, VA: University of Virginia Press, 2015), 33-34

and without a purpose. Washington observed to Benjamin Harrison that, *"We are now, in appearance, launching into a wide and boundless field – puzzled with mazes and o'erspread with difficulties"....* [60]

The Virginia Line is Ordered Southward

A bit more clarity began to appear to Washington in November when he received the first reports of British preparations to embark a large force aboard ships in New York.[61] He did not know it at the time, but General Clinton had decided to send 8,000 troops south to capture Charleston, South Carolina.

Two weeks after General Washington discovered that General Clinton might have a new operation planned, he received instructions from Congress to send the continental troops from North Carolina, as well as whatever other troops he could spare, to Charleston, South Carolina to reinforce General Benjamin Lincoln's southern army.[62] General Washington, convinced that the active campaign season for 1779 had ended in the north, was well into planning for his winter encampment in New Jersey, New York, and Connecticut when he received these instructions and initially ordered just the North Carolinians and some dragoons to march south.[63]

---

[60] Ibid., 34
[61] Ferraro, ed., "General Wayne to General Washington, November 4, 1779," *The Papers of George Washington*, Vol. 23, 154
[62] Ferraro, ed., "Samuel Huntington to General Washington, November 11, 1779," *The Papers of George Washington*, Vol. 23, 243
[63] Ferraro, ed., "General Washington Samuel Huntington, November 20, 1779," *The Papers of George Washington*, Vol. 23, 377

Upon further consideration, however, Washington grudgingly decided that the Virginian troops should proceed to South Carolina as well.[64] General Washington urged Congress to arrange water transport for the troops for as much of the journey as possible to protect them from fatigue and also reduce the chance of desertions. He waited a week to inform General Woodford of his decision, hoping that Congress might use the time to make the necessary arrangements to transport the troops south.

On December 6th, Washington revealed his decision in a letter to General Woodford, who was in Morristown, New Jersey with his troops building winter huts.

> *As it is highly probable the Virginia Troops will shortly move to the Southward, it is necessary in order that you may be prepared for such an event to give you notice of it. But as it is very much my wish to keep it secret, I must entreat you to take every necessary step to prepare them for marching without disclosing the intention.*[65]

Washington's order to march south came two days later.

> *I have this minute been honoured with a Letter from Congress...directing the Troops of the Virginia line to be put in motion immediately. You will put everything in train and march the whole, with their Tents & baggage as soon as possible to Philadelphia,*

---

[64] Ferraro, ed., "General Washington to Samuel Huntington, November 29, 1779," *The Papers of George Washington*, Vol. 23, 482
[65] Ferraro, ed., "General Washington to General Woodford, December 6, 1779," *The Papers of George Washington*, Vol. 23, 484

where you will receive farther Orders from Congress....[66]

---

[66] Ferraro, ed., "General Washington to General Woodford, December 8, 1779," *The Papers of George Washington*, Vol. 23, 559

Chapter Eleven

"We arrived…to the great joy of the garrison."

1780

Moving over 3,000 troops and six, 6 pound brass cannon hundreds of miles to the south in the onset of winter with very short notice presented a number of challenges.[1] Over a thousand of the troops were due to leave the army in the next few months when their enlistments expired and most had no intention of re-enlisting or marching to South Carolina (although they were more than happy to march back to Virginia.)[2] General Washington preferred that they not remain with the main army and left it to Congress to decide what to do with them.[3] Approximately 500 of Woodford's troops had inadequate shoes and were unable to march, an issue that was fortunately addressed with a supply of new shoes.[4] At the urging of General Washington, Congress tried to arrange water transport for General Woodford and his troops, at least for large portions of the journey, but the presence of ice in the

---

[1] Ferraro, ed., " General Washington to Samuel Huntington, December 10-11, 1779," *The Papers of George Washington*, Vol. 23, 567 and 485
[2] Ferraro, ed., *The Papers of George Washington*, Vol. 23, 485
[3] Ferraro, ed., "General Washington to General Woodford, December 24, 1779," *The Papers of George Washington*, Vol. 23, 711-712
[4] Ferraro, ed., "Tench Tilghman to John Mehelm, December 10, 1779," *The Papers of George Washington*, Vol. 23, 485

Delaware River and the absence of French warships in the Chesapeake Bay made ship transport too dangerous.[5]

Then there was the issue of command. Although General Muhlenberg had continued to command his brigade in 1779, he still disputed General Woodford's advancement over him and now refused to serve under Woodford in South Carolina. Muhlenberg appealed one last time to Congress for reconsideration, but it was to no avail, the best Congress could do was issue a resolution declaring that General Muhlenberg's loss of rank was in no way a reflection upon his conduct or character. This appeased General Muhlenberg enough to convince him to remain with the army, but he still refused to serve under General Woodford's command. Surrendering command of his brigade, the Board of War sent Muhlenberg back to Virginia to take charge of raising new troops for the continental army. With the entire Virginia continental line now under his command, General Woodford proposed yet another reorganization of the troops, forming all of those marching to Charleston with him into three battalions totally approximately 2,000 men.[6]

General Woodford commenced his march south on December 9th, staggering the departure of units over several days for logistical reasons. General Washington wrote to Woodford on the eve of the departure of Woodford's rear detachment, issuing a fond farewell to the general and his troops.

---

[5] Ferraro, ed., *The Papers of George Washington*, Vol. 23, 485
[6] Henry A. Muhlenberg, *The Life of Major-General Peter Muhlenberg of the Revolutionary* Army, (Philadelphia: Carey and Hart, 1849), 181 and Ferraro, ed., "General Washington to General Woodford, December 24, 1779," *The Papers of George Washington*, Vol. 23, 711-712

I sincerely wish you and the troops under your command a comfortable march and a speedy arrival. The interests of America may very essentially require the latter, towards which I am perswaded you will do all in your power. Nothing will make me happier than to hear at all times that the Virginia line distinguishes itself in every qualification that does honor to the military profession. Its composition is excellent – and a strict attention to discipline will always entitle it to vie with any Corps in this, or any other service.

They are going into a new & probably important field; to act with troops to whom they have been hitherto strangers – This ought to prove an additional incitement to a spirit of emulation. My affection for the troops, & my concern for the credit of the army under my command, as well as for their own credit make me anxiously desire the Officers may exert themselves to cultivate that perfection in discipline on which the usefulness & reputation of a Corps absolutely depends.... I entreat you to communicate what I have said to the Gentn of the line; and at the same time to assure them of my warmest esteem & best wishes for their welfare & success. With the truest regard I am – Dr Sir Yr most Obedt Ser.
Geo: Washington[7]

General Woodford and his troops marched from Morristown to Trenton in mid-December. A newspaper account of their arrival described them, *"in high spirits and*

---

[7] Ferraro, ed., "General Washington to General Woodford, December 13, 1779," *The Papers of George Washington*, Vol. 23, 602-603

[they] *make a martial appearance."*[8] They halted in Trenton to await the arrangements of Congress and paid their respects to Mrs. Washington on December 28th when she passed through on her way to join her husband in Morristown.[9] General Woodford had gone ahead to Philadelphia to consult with Congress on the transportation arrangements and returned to Trenton with bad news. He shared it with General Washington in a letter on December 28th.

> *After providing the necessaries for the Troops at Philadelphia, I came up to put them in motion from this place. The Board of War have determined that they March in three divisions for the conveniency of being accommodated upon the Road. The first division Marched Yesterday, the second will move Tomorrow, & the third on Friday, provided the Weather is not too bad.*
>
> *I expect the Whole will get from Philadelphia early in the next week. We are order'd the Rout of Lancaster, York, Frederick Town &ct orders have been sent to the different posts to make the necessary provisions for us, but if I am to Judge from the present state of the several departments at Philadelphia I fear their will be no certainty of being supplied. I see many difficulties in this long March by Land at this Season, but your Excellency may depend upon no time being lost that it is possible to avoid.*[10]

---

[8] Ferraro, ed., "New Jersey Gazette, December 15, 1779," *The Papers of George Washington*, Vol. 23, 485

[9] Ferraro, ed., *The Papers of George Washington*, Vol. 23, 485

[10] Ferraro, ed., "General Woodford to General Washington, December 28, 1779," *The Papers of George Washington*, Vol. 23, 768

It appears that Congress chose to send the Virginians whose enlistments were soon to expire back to General Washington in Morristown, a decision he, and the impacted troops, likely did not approve of.[11]

Woodford's troops reached Philadelphia as expected, but bad weather delayed the departure of General Woodford and the last of three detachments for over a week.[12] Woodford updated General Washington on the challenges he faced moving the troops southward.

> *The extreme badness of the weather has detain the Troops much longer here then I could have expected, togeather with the difficulty of getting them supplied with necessaries for so long a March by land; they are still deficient in Blankets, Breeches & Shirts, for other things are tolerably well off.*
>
> *The two divisions under Colos. Russell & Nevill with the Artillery, are by this time at or near Lancaster; the third & last division Marched this Day, & Tomorrow I shall leave Town to get into the Front, & after giving the necessary orders shall proceed to Fredericksburg to make provision for their reception.*
>
> *It was thought best by the Board of War to march them separately in this severe weather, for the conveniency of accommodateing them with quarters, but by the time they reach Fredericksburg I should hope it will be more moderate, that the Tents may be pitched & the Troops move*

---

[11] Ibid.
[12] Benjamin L. Huggins, ed., "General Woodford to General Washington, January 13, 1780," *The Papers of George Washington*, Vol. 24, (Charlottesville, VA: University of Virginia Press, 2016), 120

in a Boddy, which will contribute much to good order & discipline....[13]

Woodford reported that he had suffered some desertions, but not as many as he expected. He feared a decision by the Board of War to honor promises made almost a year earlier for winter furloughs for many of the troops would significantly decrease his troop strength.

> *A number of men who reinlisted for the war late in last winter & in the spring were not furloughed, but promised that indulgence this winter, which most of them claim – the Board have directed that the publick Faith should not be violated, but the engagement complied with; this will thin our Battalions, & from the great number of inducements they will meet with in the State to violate their Furloughs, I apprehend a bad account of them in the Spring. I have endeavour'd to persuade them to stay for this indulgence till a future Day, I have been able to prevail but in few instances.*[14]

Congress did take one action to significantly assist General Woodford's march; it authorized warrants amounting to $350,000 to General Woodford so that he could pay the troops what was owed them as well as the expected expenses of the long march.[15]

---

[13] Ibid.
[14] Ibid.
[15] Stewart, "J. Barrall to Joseph Clay, Deputy Pay Master General, January 29, 1780," *The Life of Brigadier General William Woodford of the American Revolution*, Vol. 2, 1143

Delayed in Virginia

General Woodford and most of his troops reached Fredericksburg, Virginia in early February. General Woodford wrote to General Washington on February 8th, from Fredericksburg with an update of his progress.

> *The first & second division of the Troops are arrived at this place; the third under command of Colo. Gist will not be here in less than five or six Days – the fatigueing March the Troops have had this extreem bad weather, the reduced situation of the waggon & artilery Horses, together with sundry repairs wanting to the waggons, has induced me to halt them here till the rear gets up – when I shall put them in motion for Petersburg.*
>
> *The continuence of the Frost will oblige me to march still by divisions; this is much against my inclination if it could be avoided, but without this method it would be imposible to procure quarters for the Officers & Men in this country.*
>
> *There has been some desertions, but not so considerable as I feard –we have picked up some Recruits & shall continue to do all we can in that way – the want of the State bounty is of great disadvantage in this business.... I understand the last assembly made no provision for recruiting their Battalions. I am much at a loss to know how the officers will be imployed, as soon as I have pass'd the Rear of the Troops through this place, I shall take Williamsburg in my way to Join them, when I Shall consult the Govr & Council upon that subject.*

> The Troops have been Healthy till lately, several have been taken Ill, & I fear the number of our sick will increase. The Men who were order'd to be furloughed by the board of war...has thin'd our Battalions & I am apprehensive we shall have a bad acct of them in the Spring.
>
> I am informed the provisions made for the Troops south of Petersburg, upon the rout pointed out by the board of war, will by no means answer the purpose of supplies -- I shall be under the necessity of altering it as circumstances may happen, provided I find it is the case.
>
> All our Friends in Fredericksburg are well, except Colo. Lewis who had been Ill some time.[16]

If Woodford was able to visit his home in Caroline County, which is very likely, his stay was brief, for the situation in South Carolina for General Lincoln and his small southern army grew more alarming each day.

General Henry Clinton, who had done little offensively in 1779 due to a lack of sufficient troops, finally initiated a bold strike against the Americans in December 1779 when he embarked over 8,000 troops aboard a fleet of nearly 90 ships in New York.[17] They sailed south, towards Georgia and ultimately, Charleston, South Carolina, which General Clinton was determined to capture. Although bad weather delayed many of his ships, most had arrived off the mouth of the Savannah River by early February.[18]

---

[16] Benjamin L. Huggins, ed., "General Woodford to General Washington, February 8, 1780," *The Papers of George Washington*, Vol. 24, 419-420

[17] Carl P. Borick, *A Gallant Defense: The Siege of Charleston, 1780*, (University of South Carolina, 2003), 23

[18] Ibid., 26

Word of the arrival of a large British fleet reached General Lincoln in Charleston by February 11th, about the same time General Clinton began disembarking his men and supplies on the coastal islands southeast of Charleston. General Lincoln immediately penned a letter to General Woodford urging him to hurry his march to Charleston.

> *The accounts I have rec'd this morning of the arrival of Sir Henry Clinton & Lord Cornwallis with a large body of Troops in Georgia – and the evidence I have of their intention being against this State – have induced me to request that you would leave your waggons & spare Baggage and hasten your March to this town. Your speedy arrival is most ardently wished for & it is not more so than necessary.* [19]

General Woodford received Lincoln's letter in Petersburg on March 6th and immediately responded, reporting that his effective force had dwindled to 737 men fit for duty.[20] He also updated General Washington, explaining the cause for his slow progress.

> *My last to your Excellency was from Fredericksburg the 8th Feby. you will no doubt be surprized that we should be near a Month in geting so short a distance, but you*

---

[19] Stewart, "General Lincoln to General Woodford, February, 11, 1780," *The Life of Brigadier General William Woodford of the American Revolution*, Vol. 2, 1152

[20] Stewart, "General Lincoln to General Washington, March 24, 1780," *The Life of Brigadier General William Woodford of the American Revolution*, Vol. 2, 1153
   Note: In this letter Lincoln refers to a letter received from Woodford dated March 6, 1780 with the information described.

> may be assured it was not possible to get the artilery & baggage on one Day sooner, & if it had not been for the assistance we recd from the Gentlemen upon the Road, they would not have reached this till the Earth was settled.
>
> The Day I arrived here I recd a letter from Genl Lincoln (a copy inclosed), this determined me to leave my Artilery, stores & baggage to follow on. The Troops march'd this morning – I took on a few Waggons to carry our Tents, as I could not think it prudent to expose the men upon so long a march. The Artilery &ct. will leave this in a few Days, Escorted by 140 Men of Colo. Bufords Regiment which is all of that corps who can be march'd at present, about 300 of them will be left, & I see no probability of their being equip'd for a march in any short time – the Colo. is now at Williamsburg upon that business. Genl Scott is here at present, but will go on without the men in a few Days.[21]

Woodford added that his force was considerably reduced by desertion, sickness, and furloughs and that much of his baggage had been damaged in the march. He lamented that

> I have done everything in my power to prevent this & other injuries to the Service, but I am concern'd to think that I shall not be able to render that service to the publick that was expected from us, or give so good an acct as I could wish of the publick stores with which my detachment was furnish'd.[22]

---

[21] Benjamin L. Huggins, ed., "General Woodford to General Washington, March 8, 1780," *The Papers of George Washington*, Vol. 24, 678-680
[22] Ibid.

Woodford closed his letter by informing Washington that the commander-in-chief's nephew, George Augustine, had joined Woodford's force as an ensign in his 2nd battalion. He then hurried southward as fast as possible.

During the period of Woodford's slow march to Petersburg, General Clinton consolidated his position along the South Carolina coast and steadily advanced towards Charleston against little opposition. Logistical challenges slowed him more than General Lincoln's troops, and Clinton's advance troops took the rest of the month to reach the northeastern end of James Island.[23] When they did so, however, they could see Charleston and its harbor.

A large part of General Lincoln's plan of defense depended on preventing the British navy from entering Charleston Harbor, and in that regard he was aided by a large sand bar at the mouth of the harbor that made navigation very tricky, especially for large, heavy warships. The British were counting on the advent of the spring flood tide, which they hoped would provide just enough extra depth to allow all except their largest ships, to cross the bar and approach the harbor.[24] Even if this occurred, however, the British still had to overcome a strong American naval presence of six frigates and nearly as many smaller vessels under Commodore Abraham Whipple.[25] Commodore Whipple had been sent to Charleston by Congress with three of the navy's eight frigates, and General Lincoln expected him to do everything in his power to defend Charleston. Unfortunately, Whipple did not

---

[23] Borick, 63
[24] Ibid., 81
[25] Ibid., 45

agree with General Lincoln's belief that defending the sandbar was critical, and he placed his ships closer to Fort Moultrie, which defended the entrance of the harbor, but not the sandbar. As a result, when the spring flood tide arrived on March 20th, the British navy was able to cross the bar. The British warships did not advance past Fort Moultrie, but numerous shallow draft flat bottomed boats were rowed past Charleston and up the Ashley River (thirteen miles) in late March to transport the British army (which had marched inland to Drayton Hall) across the Ashley River to Charleston Neck.[26] By the end of March these troops were encamped across the neck just two miles from General Lincoln's main defensive line for Charleston.[27]

While General Clinton advanced to the outskirts of Charleston, General Woodford pushed his Virginians southward from Petersburg as fast as possible. They reached Camden, South Carolina (approximately 350 miles from Petersburg) by the end of March where Woodford updated General Washington on both the status of his force and the latest reports he had received (a week old) on the activity around Charleston.

> *We arrived here* [Camden] *last evening in twenty-three days from Petersburg. I have only left thirteen sick upon the road, which an officer is bringing up. We have a few sick to leave here, the rest well and in good spirits.*

---

[26] Ibid., 103-104
[27] Ibid., 107

My artillery and stores are about five or six days march in the rear – they will halt here till Genl'l Lincoln's pleasure is known. My last letter from the General was dated the 17th but I have seen private letters here to the 25th when all was well. The Enemy go on slowly in their approaches. They have got some of their ships over the bar the last springtide and have advanced their works to a place called Waspoocutt distant from Town, one mile and a quarter. They have fortified several places, upon Ashley River which as far as I can be informed, they have entire possession. Our ships are to be sunk to obstruct the channel and their men and guns added to the Garrison.... I hope we shall still be there in time to be usefull, as we march upwards of twenty miles every day that we are not plagued with a Ferry.[28]

---

[28] Stewart, "General Woodford to General Washington, March 31, 1780," *The Life of Brigadier General William Woodford of the American Revolution*, Vol. 2, 1163-1164

Siege of Charleston

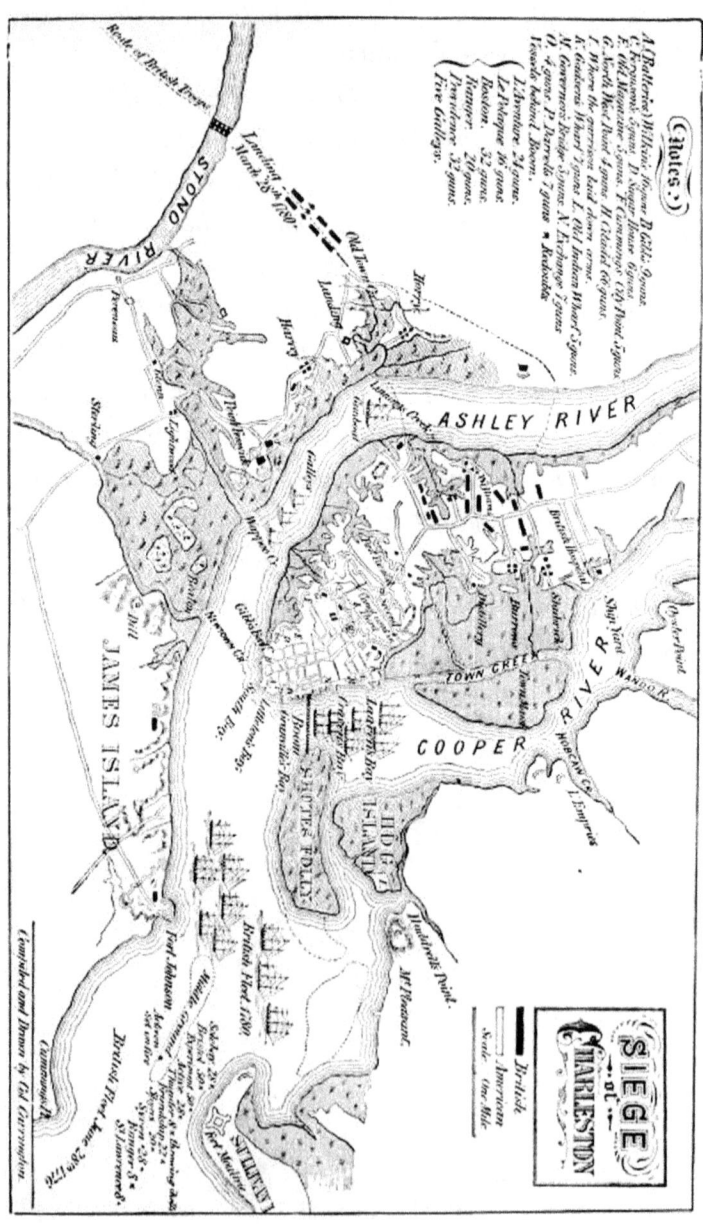

Woodford Reinforces Charleston

General Woodford and his troops, approximately 700 strong, still had 125 miles to march to reach Charleston. They covered that distance in a week and arrived in Charleston on April 7th, to great fanfare from the American army and Charleston's inhabitants. Woodford wrote to Washington to inform him of their arrival.

> *After a forced march of 505 miles, which we performed in 30 days, I had the pleasure of throwing my troops into town in good health and spirits, by the only passage now left open. We arrived on the 7th at two o' clock to the great joy of the garrison.*[29]

The joy generated by General Woodford's arrival was cut short the next day by news that the British navy had successfully sailed past Fort Moultrie with only minimal damage inflicted upon it. Although Charleston's shore batteries still protected the town from a sea borne assault, the British noose had tightened around Charleston.

Despite the fact that his artillery had yet to fire a shot at the American earthworks, General Clinton believed Lincoln's situation was untenable and summoned the American commander to surrender. Clinton's demand was delivered under a flag of truce on April 10th, and promptly dismissed by General Lincoln and his officers, all of who recognized their

---

[29] Stewart, "General Woodford to General Washington, April 9, 1780," *The Life of Brigadier General William Woodford of the American Revolution*, Vol. 2, 1165

dire situation but felt they still had options other than surrender to consider.

The day before General Clinton sent his summons, General Woodford wrote to General Washington with his assessment of the situation. Woodford was apparently ill, which limited his ability to inspect the entire defenses of the town. He also acknowledged his inexperience in siege warfare, confessing that, *"My want of experience in the defense...of a place, will not enable me to give your Excellency my opinion upon the whole of our situation, with any degree of precision."*[30] General Woodford was experienced enough to recognize the danger posed by the British navy, declaring to Washington that, *"If they can get past the obstructions in the mouth of the Cooper River, they will enfilade our works, and the consequences will be fatal, besides entirely cutting off all communication with the country."*[31] Woodford also commented on the American earthworks, noting that, *"Our batteries are strong, and the artillery numerous, but it appears to me they will not be so well manned as I could wish. Some part of the lines are rather low and I think the men off duty will be much exposed when the Batteries open."*[32] Despite these concerns, General Woodford observed that, *"The garrison appears in high spirits, and our arrival seemed to give them fresh confidence."*[33]

Lieutenant Colonel John Laurens, a former aide-de-camp to General Washington who had returned to South Carolina to help defend his home state observed that,

---

[30] Ibid, 1165-1166
[31] Ibid., 1165
[32] Ibid.
[33] Ibid.

Our obstructions in the Cooper river are completed which give a prospect of our maintaining a communication with the country....Since the arrival of General Woodford, General Lincoln will have it in his power to execute his plan of establishing the necessary posts for this purpose on the eastern shore of the river.[34]

General Lincoln had more on his mind than maintaining a communication with the country when he met with General Woodford and his other generals at a council of war on the morning of April 13th. Lincoln asked the officers for their views on the evacuation of the army from Charleston.[35]

There is no record of General Woodford's position on the question, but General Lachlan McIntosh of Georgia argued strongly in favor of evacuation and believed most of the officers at the council agreed with him.[36] Unfortunately for the Americans, discussion of this idea was cut short when British artillery batteries along their first parallel opened fire on the American lines. This marked a new phase of the siege and the American officers hastily returned to their posts to endure a twelve hour artillery barrage.

In the evening, the British, who also endured American return fire, began to dig a second parallel only 300 yards from the American line.[37] They also sent a force across the Cooper River to threaten Charleston's last communication and supply

---

[34] Stewart, "Lt. Col. John Laurens to General Washington, April 9, 1780," *The Life of Brigadier General William Woodford of the American Revolution*, Vol. 2, 1167

[35] Borick, 139

[36] Ibid.

[37] Borick, 161; and Captain Johann Hinrichs, *The Siege of Charleson: Diaries and Letters of Hessian Officers*, trans. & ed. Bernhard A. Uhlendorf, (Ann Arbor: University of Michigan Press, 1938), 249

link to the north, the very route the Americans would need if they were to attempt to escape.

Monck's Corner and Lampier's Point

Although General Clinton and the British navy had successfully severed access to Charleston from the sea, the Ashley River, and from Charleston Neck, there remained one important route to the northeast, across the Cooper River that allowed supplies and communications to reach Charleston. More importantly, it served as a possible escape route if General Lincoln chose to evacuate Charleston and save his army. Two crucial posts defended this route for the Americans. The first, Lampier's Point, was within sight of the American lines and protected the crossing point of the Cooper River. Although it was only defended by two hundred men under Colonel Francois Malmedy, a volunteer officer from France, it was fortified with strong earthworks and heavy cannon.[38] The other important American post was some 35 miles up the Cooper River at Moncks Corner. If the British were going to send troops across the Cooper River to completely seal off Charleston, General Lincoln expected them to cross at Moncks Corner. He therefore posted most of his cavalry, supported by a detachment of militia (altogether some 400 strong) to guard Moncks Corner.[39] As long as the Americans held Monck's Corner and Lamprier's Point, an avenue of escape remained, and vital supplies and reinforcements could reach Charleston.

---

[38] Borick, 144
[39] Ibid., 148

On the evening of April 13th, British Lieutenant Colonel Banastre Tarleton advanced towards Monck's Corner with a detachment of cavalry to sever this link to the north. They arrived before dawn, overwhelmed the lone American patrol posted on the road, and charged into the startled American camp. Colonel Tarleton described the attack:

> *The Americans were completely surprised... many officers and men, fled on foot to the swamps, close to their encampment, where being concealed by the darkness, they effected their escape; Four hundred horses...fell into the hands of the victors; about one hundred officers, dragoons, and hussars, together with fifty waggons, loaded with arms, clothing, and ammunition, shared the same fate.* [40]

In addition to these losses, the Americans suffered fifteen dead and eighteen wounded. Only one of Tarleton's men was killed and two others wounded.[41] Colonel Tarleton's aggressive tactics had smashed the American cavalry and opened the way for General Clinton to seal off Charleston. He ordered General Charles Cornwallis to position over 2,000 troops east of the Cooper River to oppose an American retreat from Charleston, if one occurred. Cornwallis was also ordered to interdict any supplies, reinforcements, or messages attempting to pass to or from the town. This proved difficult, because the Americans still held Lamprier's Point, and General Cornwallis

---

[40] Banastre Tarleton, *A History of the Campaigns of 1780-1781 in the Southern Provinces of North America*, (NH: AYER Company, 1999), 16 Originally printed in 1787

[41] Borick, 149 and Tarleton, 17

did not have enough men to guard all of the ground northeast of Lamprier's Point.[42]

General Lincoln, aware of the importance of Lamprier's Point, sent Lieutenant Colonel Henry Laurens with 300 men, including the army's light infantry troops, to bolster its defense.[43] These reinforcements, in conjunction with the strong earthworks of the post, dissuaded General Cornwallis from launching a direct assault on the position.[44]

In Charleston, the Americans continued to bombard British work parties and inflict casualties, but the enemy steadily crept forward. Captain Ewald described the intensity of the American fire:

> *The fire of the besieged was extraordinary: they fired scrap iron and broken glass. Although this fire is not very dangerous, and the fragments usually fly up in the air, my men lost their composure and thought of nothing else but to conceal themselves....*[45]

The British completed their second parallel in mid-April, despite the hot fire of the Americans, and immediately began work on a third. Captain Ewald noted that,

> *We have now approached so close that one could easily throw a stone into the advanced ditch on the other side, which is dressed with pointed trees. I*

---

[42] Borick, 159; 182-184
[43] Ibid., 183
[44] Ibid., 185
[45] Ewald, 231

> *occupied the heads today, and was kept warm with stone missiles and scrap iron.* [46]

The two armies were within musket range of each other, which made life in the trenches even harder. Grueling work in the heat and mud under frequent bombardment and deplorable conditions was the norm for the men on both sides. With just a few hundred yards separating the two sides, the ever-present danger of small arms fire added to the misery of trench warfare. Losses mounted in both armies.

Casualties did not alarm General Lincoln as much as the army's dwindling supplies. On April 20th, Lincoln held another war council to discuss the situation. The garrison's food supply and ordinance was running low, and there was significant doubt whether they could hold the town much longer. Some of the officers again proposed an evacuation of the town. While this was debated, a few of Charleston's civilian leaders arrived and joined the discussion. They strenuously objected to an evacuation and threatened to turn against the American army and assist the enemy if a withdrawal was attempted. These objections ended the discussion, and the council adjourned with nothing resolved.

Four days after the council meeting, General Lincoln surprised the enemy with an early morning bayonet attack by two hundred Virginia and South Carolina continentals.[47] Captain Ewald described the affair:

> *The enemy attack was made with bayonet in hand, without firing a shot. The* [British] *light infantry abandoned a part of their post and rushed back to the*

---
[46] Ibid., 232
[47] Borick, 177-178

second parallel, whereby the jagers had to pay for the feast. Since it was not yet daylight and they could not shoot, the jagers defended themselves with their hunting swords. Two jagers were bayoneted, four severely wounded, and two, along with eight Englishmen, were captured.[48]

The impact of the attack extended into the next evening when American sentries, firing into the dark, alarmed British troops in the third parallel. Afraid that another sortie was underway, the British fled to the rear towards their second parallel. The troops there were just as anxious and fired at their retreating comrades under the mistaken belief that they were Americans. The truth was that no one had left the American lines.[49]

The British were not the only ones embarrassed by false information, however. On April 27th, Colonel Malmedy, at Lamprier's Point, received news that a large enemy force was en route to attack his post. Malmedy believed that his detachment, which had been reduced to just 100 continentals and 200 militia after the light infantry was withdrawn to reinforce the main lines, was no match for the British and he hastily withdrew across the river.[50] The last avenue of escape, supply, and communication for the Americans was abandoned to the enemy without a fight and their fate was now sealed.

Three days later the British siege lines reached the first obstacle of the American line, a canal cut across the peninsula.[51] The water in it was drained and the British

---

[48] Ewald, 233
[49] Borick, 178-179
[50] Ibid., 187-189
[51] Ewald, 234

continued forward, towards the abbatis and outer wall of the American line. On May 7th more bad news befell the Americans when Fort Moultrie surrendered.[52] Although the fort's military importance had significantly diminished after the British navy slipped past it in April, Fort Moultrie remained a symbol of resistance, and its capture was another blow to the Americans.

General Clinton hoped that these developments had weakened his adversary's resolve and that General Lincoln would re-consider capitulation. On May 8th, he once again summoned the garrison to surrender. A two day cease fire ensued as letters were exchanged about the terms of a surrender. Disagreement over the status of the militia and civilians of Charleston undermined the negotiations, and a massive discharge of artillery and musket fire from both sides signaled the resumption of hostilities. Captain Johann Hinrichs of the German jaegers described the action:

> *At eight o'clock the armistice was over. The enemy rang all the bells in the city and after a threefold Hurray! Opened a cannonade more furious than any before...In the morning our guns and howitzers opened a murderous fire...The cannonade lasted the entire day, and during the night a great number of shells were thrown.*[53]

---

[52] Borick, 206
[53] Captain Hinrichs, *The Siege of Charleston*, trans & ed. Bernhard Uhlendorf, 287

Charleston Surrenders

Despite General Lincoln's determination to hold out for better terms of surrender, the will to resist among his men, (especially the militia), was largely gone. Some of the militia abandoned their post and over 500 petitioned General Lincoln to accept Clinton's terms. Even the civilian leaders of Charleston, who were so adamant about the army fighting to the last, pleaded with General Lincoln to surrender the town.[54] With supplies dwindling, morale plummeting, and hope of relief non-existent, General Lincoln decided that further resistance was futile. He informed General Clinton on May 11th that the terms of capitulation were accepted, and the Americans formally surrendered the next day.

Considering the duration of the siege and the amount of ordinance expended by each side, casualties were surprisingly light. Over 300 troops were killed or wounded on each side.[55] Many more Americans died after the surrender aboard prison barges in Charleston harbor. Over 5,000 Americans surrendered at Charleston, including the bulk of the Virginia continental line.[56]

General Woodford's Health Declines Precipitously

General Woodford and his fellow officers fared better than their soldiers in terms of their treatment. They were held in barracks built by the Americans at Haddrell's Point. Unlike many of their enlisted men, who were confined within the holds of prison hulks in Charleston Harbor, the American

---

[54] Borick, 216-217
[55] Borick, 222
[56] Ibid., 222-223

officers at Haddrell's Point were granted permission to move about the vicinity of Christ Church parish.

Unfortunately, this indulgence did little to assist General Woodford. His health deteriorated significantly during his captivity in South Carolina. No doubt the hot, humid weather contributed to his failing health and misery. His requests to be paroled to Virginia, or even New York, were initially ignored, and his condition grew steadily worse. In June, Woodford wrote a letter to Congress requesting their help.

> *My health still continues bad, and I dread the approaching season, unless I can get leave to go farther North, which there seems no prospect of, at present. I have applyed to have my parole extended to Virginia, or if that indulgence was inadmissible, to be sent to N. York or any other Northern Post in the King's possessions; this has been refused.*
>
> *I believe it is unprecedented on our side to refuse such indulgence upon the score of health to any officer of theirs. If a representation of this affair will be likely to extricate me from my present disagreeable situation, I will thank you to do what you can for me.*[57]

General Washington learned of Woodford's perilous condition in July but confessed he knew not how to remedy the situation.

---

[57] Stewart, "General Woodford to Board of War, June 2, 1780," *The Life of Brigadier General William Woodford of the American Revolution,* Vol. 2, 1179

> *I am exceedingly sorry for General Woodford's indisposition, and that he should not have been able to obtain leave to come to Virginia or move to the Northward. I do not know what can be done to procure the indulgence.*[58]

Washington suggested bringing Woodford's situation to the attention of the British commander in Charleston and urged Congress to take whatever measures they could to ease the suffering of all American prisoners held by the British.

General Woodford was made to wait until the fall before the British consented to send him to New York. In a letter to General McIntosh written on the eve of his departure from Charleston, General Woodford thanked his fellow officer for inquiring about his health and forthrightly described his condition.

> *I wish I could tell you it was better, but that is not the case. The stagnate air of this place has disagreed with my shattered frame. The Packet it is said will sail tomorrow, but I fear it will be otherwise. Delays are dangerous with me. If any unforeseen accident should stop the packet, or should I be countermanded it is all over with me, for Dr. Garden has candidly given me his opinion that my cure is out of the power of medicine, and that my only chance is the sea air, change of climate and living. I shall avail myself of any opportunity that my good fortune may give me of*

---

[58] Stewart, "General Washington to the Board of War, July 5, 1780," *The Life of Brigadier General William Woodford of the American Revolution*, Vol. 2, 1193

once more seeing his Excellency General Washington to represent your peculiar situation.[59]

General Woodford's departure from South Carolina came too late; he died in New York City on November 13, 1780, soon after reaching New York.[60] He was buried somewhere in the graveyard of Trinity Church by the British.[61] An account of Woodford's death appeared in a Pennsylvania paper a month after his passing.

We are sorry to announce to the public that a paragraph which appeared in a New York paper a few days ago, mentioning the death of the brave and worthy General Woodford turns out to be but too true. He departed this city [Philadelphia] *in January last, at the head of a number of Virginia troops, and by persevering and rapid marches in that uncommonly cold and inclement season, so far accomplished his design as to get to Charlestown with the detachment under his command a few days before the place was closely invested. The fatigues of the siege, in which he bore a very active part, together with the mortification of becoming a prisoner, and the rigorous confinement he suffered, proved too much for his delicate constitution. In the last decline of his health, he was removed from*

---

[59] Stewart, "General Woodford to General McIntosh, September 26, 1780," *The Life of Brigadier General William Woodford of the American Revolution*, Vol. 2, 1189

[60] Stewart, "British Account of Woodford's Death in Royal Gazette," *The Life of Brigadier General William Woodford of the American Revolution*, Vol. 2, 1202

[61] Ibid.

Charles Town to New York, where he in a short time paid the debt to nature, and fell a chearfull sacrifice to his country's glorious cause.[62]

Sadly, time has erased General Woodford's gravesite; he lies somewhere in the Trinity churchyard in an unmarked grave. However, General Woodford's legacy as a patriot who defended the rights and independence of his fellow Americans for five years, and who ultimately sacrificed his life to do so, remains.

---

[62] Stewart, "Pennsylvania Packet, December 16, 1780," *The Life of Brigadier General William Woodford of the American Revolution*, Vol. 2, 1187

Bibliography

Primary Sources

----------- . *The Annual Register for the Year 1776*. 4th ed.

Anderson, D.R., ed. "Colonel Woodford to Edmund Pendleton, 26 November, 1775," in "The Letters of Colonel William Woodford, Colonel Robert Howe, and General Charles Lee to Edmund Pendleton," *Richmond College Historical Papers*, June, 1915.

Boyd, Julian P., ed. *The Papers of Thomas Jefferson,* Vol. 1. Princeton, NJ: Princeton University Press, 1950.

Boyle, Joseph Lee. *Writings from the Valley Forge Encampment of the Continental Army,* Vol. 1-5. Bowie: Heritage Books Inc., 2000-05.

Campbell, Charles, ed. *The Bland Papers*, Vol. 1. 1840.

Clark, William ed. *Naval Documents of the American Revolution,* Vol. 2-3. Washington, D.C., 1964.

Cresswell, Nicholas. *The Journal of Nicholas Cresswell.* New York: The Dial Press, 1974.

Davis, K.G. ed. *Documents of the American Revolution*, Vol. 12. Irish University Press, 1976.

Dorman, John ed., "Eden Clevenger Pension Application," *Virginia Revolutionary Pension Applications*, Vol. 20.

Ewald, Johann. *Diary of the American War: A Hessian Journal*, (New Haven & London: Yale University Press, 1979
  Translated & edited by Joseph P. Tustin

Force, Peter, ed., *American Archives, Fourth Series,* Vol. 4. Washington, D.C., U.S. Congress, 1858-1853.

Ford, Worthington C. ed. *Journals of the Continental Congress: 1774-1789,* Vol. 1-7 Washington, DC.: Government Printing Office, 1904-1908

Hamilton, Stanislaus M. ed. *Letters to Washington & Accompanying Papers*, Vol. 5. Boston & New York: Houghton Mifflin, Co., 1902.

Hening, William W., ed. *The Statutes at Large Being a Collection of all the Laws of Virginia,* Vol. 9. Richmond: J. & G. Cochran, 1821.

Heth, William. "Orderly Book of Major William Heth of the Third (sic) Virginia Regiment, May 15-July 1, 1777 *Virginia Historical Society Collections,* New Series 11, 1892.

Hinrichs, Captain Johann *The Siege of Charleson: Diaries and Letters of Hessian Officers.* Ann Arbor: University of Michigan Press, 1938.
  Translated and edited by Bernard A. Uhlendorf

Journal of the House of Delegates, 1835-36, Doc. No. 43. Richmond, 1835, Virginia State Library.

Lee Papers, Vol. 1. Collections of the New York Historical Society, 1871.

Lesser, Charles H., ed. *The Sinews of Independence, Monthly Strength Reports of the Continental Army.* University of Chicago Press, 1976.

Mayes, David John, ed. *The Letters and Papers of Edmund Pendleton,* Vol. 1. Charlottesville: University Press of Virginia, 1967.

McIlwaine, H.R. ed. "Proceedings of the Committees of Safety of Caroline and Southampton Counties, Virginia: 1774-1776, *Bulletin of the Virginia State Library,* Vol. 17, No. 3. November, 1929.

Moore, Frank, ed., "Extract of a letter from an officer at Paramus," *Diary of the American Revolution,* Vol. 2.

Reese, George, ed. *The Official Papers of Francis Fauquier, Lt. Governor of Virginia, 1758-1760,* Vol. 1-3. Charlottesville: University Press of Virginia, 1980.

Rutland, Robert A. ed.*he Papers of George Mason,* Vol. 1. University of North Carolina Press, 1970.

Schreeven William Van, and Robert L. Scribner, eds. *Revolutionary Virginia: The Road to Independence,* Vol. 1-2. University Press of Virginia, 1973-75

Scribner, Robert L, ed. *Revolutionary Virginia: The Road to Independence* Vol. 3. University Press of Virginia, 1977.

Scribner Robert L. and Brent Tarter, eds. *Revolutionary Virginia: The Road to Independence,* Vol. 4-7. University Press of Virginia, 1978-83.

Smith, Paul H. *The Letters of Delegates to Congress*, Vol. 1. Library of Congress, 1976.

Sparacio, Ruth and Sam. eds. *Virginia County Court Records Order Book, Caroline County, Virginia, 1765.* The Antient Press, 1989.

Stewart, Catesby Willis, ed. *Woodford Letter Book: 1723-1737.* Verona, VA: McClure Printing Company, Inc., 1977.

Tarleton, Banastre. *A History of the Campaigns of 1780-1781 in the Southern Provinces of North America.* NH: AYER Company, 1999.
    Originally printed in 1787

Tarter, Brent, ed."The Orderly Book of the 2nd Virginia Regiment," *Virginia Magazine of History and Biography*, Vol. 85, No. 2 April 1977.

Weedon, George. "Brigadier General George Weedon's Correspondence Account of the Battle of Brandywine, September 11, 1777," Manuscript in the Chicago Historical Society. Transcribed by Bob McDonald

Papers of George Washington, Colonial Series

Abbot, W.W. ed. *The Papers of George Washington, Colonial Series*, Vol. 3-6. Charlottesville: University Press of Virginia, 1984-1988.

Abbot, W.W. and Dorothy Twohig, eds. *The Papers of George Washington: Colonial Series,* Vol. 7. Charlottesville: University Press of Virginia, 1990.

Papers of George Washington, Revolutionary War Series

Chase, Philander D., ed. *The Papers of George Washington, Revolutionary War Series,* Vol. 1-5, 7. Charlottesville, VA: University Press of Virginia, 1985-97

Chase, Philander D. and Frank Grizzard Jr., *The Papers of George Washington, Revolutionary War Series,* Vol. 6. Charlottesville, VA: University Press of Virginia, 1994.

Grizzard, Frank, Jr., *The Papers of George Washington, Revolutionary War Series,* Vol. 8. Charlottesville, VA: University Press of Virginia, 1998.

Chase, Philander D. ed. *The Papers of George Washington, Revolutionary War Series,* Vol. 9. Charlottesville, VA: University Press of Virginia, 1999.

Gizzard, Frank W. Jr., ed. *The Papers of George Washington, Revolutionary War Series,* Vol. 10. Charlottesville, VA: University Press of Virginia, 2000.

Chase, Philander D. and Edward G. Lengel, eds. *The Papers of George Washington, Revolutionary War Series,* Vol. 11. Charlottesville, VA: University of Virginia Press, 2001

Grizzard, Frank E. Jr. and David R. Hoth, eds. *The Papers of George Washington, Revolutionary War Series,* Vol. 12. Charlottesville, VA: University of Virginia Press, 2002.

Lengel, Edward G. ed. *The Papers of George Washington, Revolutionary War Series,* Vol. 13. Charlottesville, VA: University of Virginia Press, 2003.

Hoth, David R. ed. *The Papers of George Washington, Revolutionary War Series,* Vol. 14, Charlottesville, VA: University of Virginia Press, 2004.

Lengel, Edward G. ed., *The Papers of George Washington, Revolutionary War Series,* Vol. 15, (Charlottesville, VA: University of Virginia Press, 2006.

Hoth, David R. ed. *The Papers of George Washington, Revolutionary War Series,* Vol, 16, (Charlottesville, VA: University of Virginia Press, 2006.

Chase, Philander D. ed., *The Papers of George Washington, Revolutionary War Series,* Vol. 17, Charlottesville, VA: University of Virginia Press, 2008.

Lengel, Edward G. Lengel, ed., *The Papers of George Washington, Revolutionary War Series,* Vol. 18, Charlottesville, VA: University of Virginia Press, 2008.

Chase, Philander D. and William M. Ferraro, eds. *The Papers of George Washington, Revolutionary War Series,* Vol. 19, Charlottesville, VA: University of Virginia Press, 2009.

Lengel, Edward G. ed. *The Papers of George Washington, Revolutionary War Series,* Vol. 20, Charlottesville, VA: University of Virginia Press, 2010.

Ferraro, William M. ed., *The Papers of George Washington, Revolutionary War Series,* Vol. 21. Charlottesville, VA: University of Virginia Press, 2012.

Huggins, Benjamin L. ed., *The Papers of George Washington, Revolutionary War Series,* Vol. 22. Charlottesville, VA: University of Virginia Press, 2013.

Ferraro, William M. ed. *The Papers of George Washington, Revolutionary War Series,* Vol. 23. Charlottesville, VA: University of Virginia Press, 2015.

Huggins, Benjamin L. ed., *The Papers of George Washington, Revolutionary War Series,* Vol. 24. Charlottesville, VA: University of Virginia Press, 2016.

"A General Return of the 12 Virginia Battalions in Morristown, May 17, 1777," *The Papers of George Washington, Revolutionary War Series,* Library of Congress online.

Writings of George Washington

Fitzpatrick, John C. ed. *Writings of George Washington,* Vols. 1-20. Washington, DC: U.S. Govt. Printing Officer, 1931-37.

Newspapers

Dixon and Hunter, *Virginia Gazette.* October 28, 1775.
Dixon and Hunter, *Virginia Gazette.* January 13, 1776.
Dixon and Hunter, *Virginia Gazette.* May 25, 1776.
Dixon and Hunter, *Virginia Gazette,* July 10, 1778.
Dixon and Hunter, *Virginia Gazette .* November 6, 13, 27, 1778.

Dixon and Hunter, *Virginia Gazette.* December 4, 1778.

Dixon and Nicolson, *Virginia Gazette.* February 12, 19, 26 1779.

Dixon and Nicolson, *Virginia Gazette.* March 5, 12, 1779

Pinkney, *Virginia Gazette.* October 26, 1775.
Pinkney, *Virginia Gazette.* November 2, 1775.

Purdie, *Virginia Gazette.* May 12, 1775, Supplement.
Purdie, *Virginia Gazette.* May 19, 1775, Supplement.
Purdie, *Virginia Gazette.* August 11, 1775.
Purdie, *Virginia Gazette.* September 22, 1775.
Purdie, *Virginia Gazette.* February 2, 1776.
Purdie, *Virginia Gazette.* March 1, 1776.
Purdie, *Virginia Gazette.* July 5, 1776.
Purdie, *Virginia Gazette*, Aug. 9, 1776.
Purdie *Virginia Gazette*, June 19, 1778

Purdie and Dixon, *Virginia Gazette*, Supplement. July, 28 1774

Rind. *Virginia Gazette.* January 17, 1771
Rind *Virginia Gazette.* June 16, 1774.

Secondary Sources

Boatner, Mark M. *Encyclopedia of the American Revolution.* NY: D. McKay Co., 1966.

Borick, Carl P. *A Gallant Defense: The Siege of Charleston, 1780.* University of South Carolina, 2003.

Braisted, Todd W. *Grand Forage, 1778: The Battleground Around New York City.* Westholme: Yardley, PA, 2016.

Cecere, Michael. *A Good and Valuable Officer: Daniel Morgan in the Revolutionary War.*, Heritage Books, 2016.

Cecere, Michael. *An Officer of Very Extraordinary Merit: Charles Porterfield and the American War for Independence, 1775-1780.* Heritage Books, 2004.

Cecere, Michael. *Captain Thomas Posey and the 7th Virginia Regiment.* Heritage Books, 2005

Cecere, Michael. *Cast Off the British Yoke: The Old Dominion and American Independence, 1763-1776.* Heritage Books, 2014.

Cecere, Michael. *Second to No Man But the Commander in Chief. Hugh Mercer, American Patriot.* Heritage Books, 2015

Cecere, Michael. *They Are Indeed a Very Useful Corps: American Riflemen in the Revolutionary War,* Heritage Books, 2006.

Cecere, Michael. *They Behaved Like Soldiers, Captain John Chilton and the Third Virginia Regiment, 1775-1778.* Heritage Books, 2004.

Cecere, Michael. *Wedded to My Sword: The Revolutionary War Service of Light Horse Harry Lee.* Heritage Books, 2012

Cunningham, John T. *The Uncertain Revolution: Washington & the Continental Army at Morristown.* Cormorant Publishing, 2007

Fischer, David Hackett. *Washington's Crossing.* Oxford University Press, 2004.

Golway, Terry. *Washington's General: Nathanael Greene and the Triumph of the American Revolution.* New York: Henry Holt and Company, 2006.

Graham, James. *The Life of General Daniel Morgan of the Virginia Line of the Army of the United States.* Bloomingburg, NY: Zebrowski Historical Services Publishing Company, 1993.
    Originally published in 1856.

Harris, Michael C. *Brandywine: A Military History of the Battle that Lost Philadelphia but Saved America, September 11, 1777.* Savas Beatie, 2014.

Hartmann, John W. *The American Partisan: Henry Lee and the Struggle for Independence: 1776-1780.* Burd St. Press, 2000.

Higginbotham, Don. *Daniel Morgan: Revolutionary Rifleman.* Chapel Hill: University of North Carolina Press, 1961.

Jackson, John W. *Valley Forge: Pinnacle of Courage.* Thomas Publications, 1992.

Lee, Henry. *The Revolutionary War Memoirs of General Henry Lee,* New York: Da Capo Press, 1979.

Lender, Mark Edward and Garry Wheeler Stone. *Fatal Sunday: George Washington, the Monmouth Campaign, and the Politics of Battle*. University of Oklahoma Press, Norman, OK, 2016.

Loprieno, Don. *The Enterprise in Contemplation: The Midnight Assault of Stony Point*. Heritage Books, 2004

Marshall, John. *The Life of George Washington,* Vol. 2. Fredericksburg, VA: The Citizens Guild of Washington's Boyhood Home, 1926.

Mayer, Henry. *A Son of Thunder: Patrick Henry and the American Republic*. Charlottesville: University Press of Virginia, 1991.

McGuire, Thomas. *The Philadelphia Campaign: Brandywine and the Fall of Philadelphia,* Vol. 1. Stackpole Books, 2006.

McGuire, Thomas. *The Philadelphia Campaign: Germantown and the Roads to Valley Forge,* Vol. 2. Stackpole Books, 2007.

Muhlenberg, Henry A. *The Life of Major-General Peter Muhlenberg of the Revolutionary Army*. Philadelphia: Carey and Hart, 1849.

Reed, John F. *Campaign to Valley Forge: July 1, 1777-December 19, 1777*. Pioneer Press, 1965.

Reed, William B. *Life and Correspondence of Joseph Reed*, Vol. 2. Lindsay and Blakiston: Philadelphia, 1847.

Selby, John. *The Revolution in Virginia, 1775-1783.* Colonial Williamsburg Foundation, 1988.

Stewart, Catesby Willis. *The Life of Brigadier General William Woodford of the American Revolution,* Vol. 1-2. Richmond, VA: Whitten & Shepperson, 1973.

Stryker, William. *The Battle of Monmouth.* Princeton: Princeton University Press, 1927.

Trussell, John B.B.Jr., *Birthplace of an Army: A Study of the Valley Forge Encampment.* Pennsylvania Historical and Museum Commission, 1998.

Ward, Harry M. *Charles Scott: Spirit of '76.* Charlottesville: University Press of Virginia, 1988.

Ward, Harry M. *Duty, Honor or Country: General George Weedon and the American Revolution.* Philadelphia, PA: American Philosophical Society, 1979.

Williams, Tony. *Hurricane of Independence:The Untold Story of the Deadly Storm at the Deciding Moment of the American Revolution.* Sourcebooks Inc., 2008.

Wirt, William. *Sketches in the Life and Character of Patrick Henry.* Philadelphia, 1817.

Wrike, Peter Jennings. *The Governor's Island: Gwynn's Island, Virginia, During the Revolution.* The Gwynn's Island Museum, 1993.

Index

3rd Continental Light Dragoons, 181
14th Regiment of Foot, 51. 58, 60, 67, 72, 76, 78-79, 81, 114

A

Acuaquenunk Bridge, NJ, 181
Adams, John, 20, 26
Alexander, Gen. William (Lord Stirling), 134, 141-143, 146-147, 170, 180, 183, 197, 201, 206-207
Arnold, Benedict, 131-132, 160
Augusta Co., VA, 131

B

Batchelors Mill, VA, 103-104
Batut, Lt. John, 66, 76, 79
Baylor, Col. George, 181
Bellow, Capt. Henry, 92-94
Bethleham, PA, 151-152
Birmingham Meeting House, 142-146
Blackburn, Adj. Thomas, 73
Bland, Richard, 19
Boston, MA, 16-17, 21, 37, 54, 131
Boston Tea Party, 16
Botetourt, Governor, 14
Boundbrook, NJ, 128
Bowling Green, VA, 29
Boykin, Lt. Francis, 103-105
Braddock, Gen. Edward, 4
Brandywine, Battle of, 140-149
Buford, Abraham, Capt. 45-46, 48 Col. 222
Bullit, Col. Thomas, 38
Burgoyne, Gen. John, 136-137, 139, 153
Burwell's Landing, 49, 109

Byrd, Col. William, 9

C

Cadwalader, Gen. John, 155
Camden, SC, 224-225
Canada, 10, 136-137
Caroline County, VA, 1, 3, 5, 11-12, 16-19, 23, 25-26, 29-32, 35, 101, 106, 119, 185, 220
Caroline Co. Militia, 3
Caroline Co. Independent Militia, 25
Cary, Col., 45
Chad's Ford, PA, 140-141
Charleston, SC, 16, 113, 139, 209, 214, 220, 223-224
Charleston, Siege of (1780), 227-237, 240
Cherokee Indians, 10
Chester, NY, 137
Chilton, Capt. John, 138
Christian, William, 34
Clark, Maj. John, 199, 202-204
Clarkstown, NY, 180-182
Clinton, Gen. Henry, 113, 168, 172, 176-177, 180, 182-183, 190, 193, 208-209, 220-221, 223-225, 227, 230-231, 235-236
Cobham, VA, 50
Continental Association of 1774, 20, 25
Continental Congress, 18-21, 24, 26, 30, 32, 51, 88, 106, 115-116, 123, 125-126, 158-164, 174, 186, 211, 213-214, 216-221, 237-238
Cornwallis, Gen. Charles, 232
Cooches Bridge, DE, 140
Culpeper Minute Battalion, 39, 45-46, 50, 56-57, 62, 73-74, 77, 84
Cumberland County, VA, 34

D

Danbury, CT, 180
Dartmouth, Lord, 69, 90
Davis, Capt., 103
Detroit, MI, 10
Dickinson, Gen. Philemon, 170
Dilworth, PA, 142
Drayton Hall, SC, 224

Dunmore, Gov. John Murray, Earl of, 17, 23, 28-29, 31-32, 42, 44, 49, 51-55, 58, 63-67, 69-70, 72, 78-80, 83, 89-91, 94-101, 109-110, 113-115, 130
Dunmore, 91

E

East India Company, 16
Eilbeck, 91
Elizabethtown, NJ, 180
Estang, Count, 190
Ewald, Capt. Johann, 144-145, 232-234

F

Fairfax County, VA, 13-14
Fairfax Independent Militia Company, 24
Fauquier, Gov. Francis, 12
Febiger, Col. Christian, 132, 178-179
Fishkill, NY, 180
Flora, Billy, 71
Forbes, Gen. John, 7-8
Fordyce, Capt. Charles, 72, 74-75, 78
Fort Bedford, 10
Fort Chiswell, 10
Fort Cumberland, 7
Fort Duquesne, 7-9
Fort George, 6
Fort Ligonier, 9-10
Fort Loudoun, 4, 9
Fort Moultrie, 224, 227, 235
Fort Necessity, 4
Fort Niagara, 10
Fort Pitt, 9
Fort Washington, 131
Frederick Co., VA, 131
Fredericksburg, NY, 180
Fredericksburg, VA, 1, 18, 28-30, 33, 106, 164, 166, 189, 217, 219-220, 222
French Alliance, (1778), 168
French & Indian War, 3-10, 115

G

Gardner Doctor, 239
Gist, Col., 203-204
Germantown, battle of, 153
Gloucester Co., VA, 45, 109

Gosport, VA, 97
Grayson, Col. William, 178
Grant, Maj. James, 8
Great Bridge, VA, 57, 86, 88-89, 100-101, 103-104
Great Bridge, battle of, 57-81
Greene, Gen. Nathanael, 129, 140, 173-174, 176, 194, 204
Gwynn's Island, VA, 114-115, 118, 130

H

Hackensack, NJ, 187, 197
Haddrell's Point, SC, 237
Hagan, Lt., 144
Hardy, Capt. Levin, 198
Hampton, VA, 39, 52
  battle of, (1775), 41-45
Hancock, John, 139
Hanover Co., VA, 26
Hanover Co. Militia, 30
HanoverTown, VA, 38
Harlem Heights, NY, 130
Harrison, Benjamin, 19, 30, 208-209
Head of Elk, MD, 140

Henry, Patrick, 19-20, 26-27, 30, 33-34, 36, 38-39, 83-88, 101, 107, 118, 120
Heth, Maj. William, 132, 179
Hinrichs, Capt. Johann, 235-236
House of Burgesses, 10-12, 14, 17, 23, 32, 197
Howe, Gen. Robert, 88-89, 91, 93-101
Howe, Gen. William, 54, 58, 124, 134-142, 144-145, 147, 153, 156, 168
Huntington, Gen., 170

I

Intolerable Acts, 16-17, 20-21
Innes, Lt. Col. James, 130

J

James Island, SC, 223
Jamestown, VA, 49, 52, 109
Jefferson, Thomas, 45, 55

K

Kemps Landing, VA, 53, 100, 104, 111
King's Ferry, NY, 193-194
King Fisher, HMS, 50-51, 89-90
King George III, 16-17
King William Co. Militia, 30

L

LaFayette, Gen. Marquis, 151-153, 170-171
Lampier's Point, SC, 230-232, 234
Lancaster, PA, 216
Laurens, Henry, 159, 229, 232
Lee, Gen. Charles, 108-114, 169-174, 177, 179
Lee, Maj. Henry, 145-146, 197-205
Lee, Richard Henry, 19-20, 24, 127-128
Leslie, Lt. Peter, 77, 80
Leslie, Capt. Samuel, 66, 69-70, 72, 76, 78-79, 81, 83

Lexington & Concord, battle of, 31, 49
Lewis, Gen. Andrew, 107, 114-115, 118-119
Lewis, Col., 220
Lewis, John, 185
Lincoln, Gen. Benjamin, 191, 210, 220-225, 227, 229-230, 232-236
Liverpool, HMS, 90, 92-93
Loudoun County, Independent Militia, 24
Lynn, Capt., 42-44, 48

M

Magdalen, HMS, 28
Malmedy, Col. Francois, 230, 234
Marshall, Lt. John, 73
Marshall, Maj. Thomas, 66, 130, 143-145
Mason, Col. David, 132
Mason, George, 14, 24
Maxwell, Gen. William, 140-141, 170, 180,
McAllister, Lt., 199
McClanahan, Col., Alexander, 13
McIntosh, Gen. Lachlan, 229, 238

McKenzie, Capt. Robert, 6
Meade, Maj. Richard, 105
Mease, James, 154
Mercer, Gen. Hugh, 33-34, 115-116, 118, 125, 130, 159
Middlebrook, NJ, 132-133, 187, 190
Minutemen, 36
Monck's Corner, SC, 230-231
Monmouth, battle of, 172-178
Monmouth Courthouse, NJ, 172-174, 176-177, 188
Montresor, Capt., John, 147
Morgan, Col., Daniel, 131-134, 156-157, 170, 179, 184, 187-188, 192, 207-208
Morris, Gen., 195
Morristown, NJ, 124, 128, 130, 136-137, 210, 215-216
Muhlenberg, Gen. John Peter, 126-127, 129, 140, 145, 149, 159-162, 164, 179, 188-189, 191, 202, 214

N

Nash, Gen. Francis, 141
Nelson, Jr., Thomas, 33-35
Nevell, Col., John, 207
New Bridge, NJ, 201
New Brunswick, NJ, 133-134
New Kent Co. Militia, 30
New Jersey, College of, 190
New Windsor, NY, 195
New York, NY, 16, 168, 239
Newark, NJ, 183
Nicholas, Capt. George, 42-44, 46, 48
Nicholas, Robert Carter, 26, 42
Norfolk, VA, 33, 39, 41, 50, 52-53, 55, 57-58, 64, 67, 80-81, 84, 89, 104-105, 110, 113,
  burned, 93-101
North Carolina Brigade, 170
Northern Neck, VA, 13

O

Osborne Hill, 142, 144
Otter HMS, 41, 51, 67, 90

P

Page, John, 45-47, 49, 55
Paramus, NJ, 181-182
Payton, Capt. Henry, 200
Peekskill, NY, 180
Pendleton, Edmund,
   criticized Boston Tea
   Party, 17
   delegate to 1st Cont.
   Congress, 19
   serves on committee to
   enforce Association, 23
   opposes P. Henry's at 2nd
   VA Convention, 30
   joins 3rd VA Conv., 34
   presides over VA Comm.
   of Safety, 38
   letter from Woodford, 63
   supports Woodford, 105-106
   at 5th VA Conv., 115
   letters to Woodford, 119-120, 122, 135-136, 165, 167, 205
Perrine Ridge, NJ, 174, 176
Perth Amboy, NJ, 134
Petersburg, VA, 220-221, 223-225
Philadelphia, PA, 16, 18-19, 30-31, 35, 107, 123, 125, 133, 138, 140, 153-155, 168-169, 177, 211, 216-217, 240
Pitt, William, 9
Poor, Gen. Enoch, 169
Port Royal, VA, 1, 25
Porterfield, Charles, 132
Portsmouth, VA, 33, 57, 101, 104, 112
Powles Hook, battle of, 196-201
Prince William Co.
   Independent Militia, 24
Princess Anne County, VA, 53, 110, 113
Princeton, NJ, 190, 194-195
   battle of, 124-125, 130
Putnam, Gen. Israel, 179-180

Q

Quebec, 10, 131
Queens Own Loyal Virginia Regiment, 55

R

Raleigh Tavern, 107
Randolph, Peyton, 18-20, 28-30
Reed, Joseph, 203
Richmond, VA, 26, 31-32
Richmond Co. VA, 106
Roebuck, HMS, 101

S

Saratoga, battle of, 153
Saunderson, Mr., 194
Scott, Charles
　Lt. Col., 35, 57-59, 61-66, 119
　General, 129, 142, 146-148, 159-160, 164, 169, 171, 179-180, 191-192, 222
Shawnee Indians, 107
Slave Trade Ban, 19
Sleepy Creek, VA, 4

Spotswood, Major Alexander, 35, 58, 71-73, 77
Spotsylvania County, VA, 35, 50
Spotsylvania Co. Independent Militia, 24
Sprowles, Andrew, 97
Squire, Capt. Mathew, 41, 43, 46, 49, 52, 90
Stamp Act, 12-14
Steenbergen, Lt. Peter, 6-7
Stephen, Gen. Adam, 116-117, 122, 129, 141-142, 146-148, 153, 159
Steuben, Gen. Freidrich von, 168
Stevens, Col. Edward, 62, 77
Stewart, Capt. Robert, 8
St. Mary's Parish, 12
Stony Point, battle of, 193, 195-196, 207, 209
Suffern, NY, 196
Suffolk, VA, 57, 59, 100-101, 112
Sullivan, Gen. John, 141, 143, 146-152, 193
Sutherland, Maj. Nicholas, 199

T

Taliaferro, John, 166, 185
Tarleton, Lt. Col.,
  Banastre, 231
Taylor, James, 19, 26
Thornton, John, 165-166,
  185, 188
Tea Act, 16
Tennent's Meeting House,
  173
Townshend Duties, 14-15
Travis, Lt. Edward, 72-73
Trenton, NJ, 215-216
  battle of, 124-125, 130
Trinity Church, NY, 239-
  240
Tucker's Point, VA, 101,
  113

V

Valley Forge, PA, 157-
  158, 160, 162-165, 167,
  169
Varnum, Gen. James, 158,
  170
Verplanck's Point, NY,
  193, 195, 209
Virginia Committee of
  Safety, 38-39, 50-52, 59-
  60, 63-64, 66, 81, 83-85,
  87-88, 105, 107
Virginia Continental
Virginia Continental
Regiments
  1st VA, 33-34, 36, 83-84,
    101, 103, 109, 119-
    120,
  2nd VA, 34-35, 38, 42,
    50, 57, 83-84, 88, 101,
    115, 118-121, 129,
    178-179, 192
  3rd VA, 115, 128, 130,
    137, 143-146, 149,
    179, 192
  4th VA, 116, 191-192
  5th VA, 109, 179 192
  6th VA, 109, 178-179
  7th VA, 109, 128, 130,
    179, 188, 192
  8th VA, 113, 191,
  11th VA, 128, 131, 179,
    188, 192
  15th VA, 128, 132, 179
Virginia Convention
  1^{s}, 18-19
  2nd, 26-28
  3rd, 30, 32-36, 38, 116
  4th, 63-64, 86, 88, 100-
    101, 104-106, 113
  5th, 115

Virginia Regiment (F & I War), 5-10

W

Waggener, Thomas, 5-6
Washington, George Augustine, 223
Washington, John Augustine, 123
Washington, Martha, 216
Washington, Samuel, 185
Washington, George,
  F & I War, 4-9
  opposes Stamp Act, 13
  proposes non-importation association plan, 14
  delegate to 1st Cont. Congress, 19
  commands Fairfax Co. Independent Militia, 24
  commands Cont. Army, 33
  approves of Woodford's appointment, 37
  Gen. Lee writes to, 108
  Woodford writes to, 116-118
  in New Jersey, 122, 124
  victories at Trenton and Princeton, 125
  writes to Woodford, 126-127
  requests Daniel Morgan command a regt., 131
  expresses concern about condition of the army, 132-133
  forms rifle corps, 133
  confused by British inactivity, 134-137, 139
  marches to confront Gen. Howe, 140
  at battle of Brandywine, 140-142
  requests to oversee troops at Bethleham, 151
  resists giving LaFayette a command, 151
  updated on Woodford's recovery, 152-153
  reacts to loss of Philadelphia, 153
  supports Woodford about clothing, 154
  remains in Whitemarsh, 155
  battle of Whitemarsh, 156
  at Valley Forge, 157-158
  embroiled in rank controversies, 159-164

instructs Woodford to recruit, 165
estimates troop strength, 167
seeks advice, 168
battle of Monmouth, 169-177
Gen. Lee court martial, 177
camps near White Plains, 178-180
letters from Woodford, 183, 186-187, 189
letter to Woodford, 188
sends reinforcements South, 191
reorganizes army, 192-193
attack on Stony Point, 195-196
atttack on Powles Hook, 197
supports Maj. Lee, 202-204
perplexed by Gen. Clinton, 208
orders VA line south to Charleston, 209-211
march arrangements, 213-214
letter to Woodford, 214-215
letters from Woodford, 217, 225, 228
laments inability to help Woodford, 238
Wayne, Gen. Anthony, 140-141, 196, 207
Weedon, Gen. George, 126-127, 129-130, 140, 143
West Point, NY, 180, 182, 193-196, 206-20849, 159-162, 164-165, 170, 179
Westmoreland Co., VA, 123
Westmorcland Co. Resolves, 13
Whipple, Commodore, Abraham, 224
White Plains, NY, 130, 178-180
Whitemarsh, PA, 153-156
William & Mary College of, 38
Williamsburg, VA, 9-11, 28-30, 33, 38-39, 45, 49, 64, 81, 86, 100-101, 108-109, 111, 114-115, 119, 219, 222
Williamsburg Powder Incident, 28

Wilmington, DE, 140
Winchester, VA, 4-5, 7, 9-10
Windsor, 1, 3, 11, 37, 106, 165, 185, 194
Witherspoon, Dr. John, 190
Woodford, Anne Cocke, 3
Woodford, Capt., Henry, 205
Woodford, John Thornton, 11, 190, 194-195
Woodford, Mary Thornton, 11, 135, 194
Woodford, Maj. William, 1, 3
Woodford, William Catesby, 11, 194
Woodford,. William, born, 1
death of father, 3
in F & I War, 4-10
returns to Windsor, 11
elected to vestry, 12
opposes Stamp Act, 13
supports non-Importation Association, 15
friends with Edmund Pendleton, 17
supports Caroline Co. Resolves, 18
serves on committee to enforce Association, 23
commands Caroline Co. Independent Militia, 25, 29-29
serves n committee of Inspection, 25
delegate to 3rd VA Convention, 31-34
appointed colonel of 2nd VA Regt., 34
writes to Gen. Washington, 37
assumes command of 2nd VA Regt., 38
ordered to Norfolk, 39
at Battle of Hampton, 45-49
crosses James River, 50, 52, 55-56
in Suffolk, 57-58
at Great Bridge, 59-81
  battle of, 69-81
dispute with Patrick Henry, 83-88
at Norfolk, 88-92
burns Norfolk, 93-100
requests furlough, 101
dispute with Lt. Boykin, 103-106
rest at Windsor, 106

returns to army, 110
commands 2nd VA Regt.,
in Suffolk, 110-115
passed over for promotion, 115-118
at Gwynn's Island, 118
trouble with his regt., 119-120
critical of P. Henry, 121
resigns his commission, 122
visits Washington, 122-123
promoted to brigadier-general, 125-128
commands 3rd VA Brigade, 128-135
thanks Pendleton, 135-136
summer 1777 operations, 137-140
at battle of Brandywine, 140-149
wounded, 149
recovers in Bethleham, PA, 151-153
rejoins his brigade, 153-156
at Valley Forge, 157-158
rank controversy, 158-164
furlough, 164-166
returns to army, 167
advises Washington, 168-171
at battle of Monmouth, 172-177
brigade restructure, 178-179
guards West Point, 180-183
furlough, 183-190
returns to army, 190
brigade restructure, 192
writes to son, 194-195
posted in New York, 195-196, 206-208
Powles Hook controversy, 202-205
writes to Pendleton, 205-206
ordered South, 210-211
marches South, 213-227
siege of Charleston, 227-236
captured, 237
ill health, 237-239
death, 239
Wright, Lt. 47-48

Y

York, PA, 216
Yorktown, VA, 33, 109

www.ingramcontent.com/pod-product-compliance
Lightning Source LLC
Chambersburg PA
CBHW062006220426
43662CB00010B/1250